Lecture Notes in Artificial Intell

Subseries of Lecture Notes in Computer Sci
Edited by J. G. Carbonell and J. Siekmann

Lecture Notes in Computer Science

Edited by G. Goos, J. Hartmanis, and J. van Leeuwen

Christian Schulte

Programming Constraint Services

High-Level Programming
of Standard and New Constraint Services

 Springer

Series Editors

Jaime G. Carbonell, Carnegie Mellon University, Pittsburgh, PA, USA
Jörg Siekmann, University of Saarland, Saarbrücken, Germany

Author

Christian Schulte
Saarland University, Programming Systems Laboratory
Postfach 15 11 50, 66041 Saarbrücken, Germany
E-mail: schulte@ps.uni-sb.de

Cataloging-in-Publication Data applied for

Die Deutsche Bibliothek - CIP-Einheitsaufnahme

Schulte, Christian:
Programming constraint services : high level programming of standard and new
constraint services / Christian Schulte. - Berlin ; Heidelberg ; New York ;
Barcelona ; Hong Kong ; London ; Milan ; Paris ; Tokyo : Springer, 2002
    (Lecture notes in computer science ; Vol. 2302 : Lecture notes in
    artificial intelligence)
    ISBN 3-540-43371-6

CR Subject Classification (1998): I.2.3, I.2.8, D.3.3, D.1.6, F.2.2, G.1.6

ISSN 0302-9743
ISBN 3-540-43371-6 Springer-Verlag Berlin Heidelberg New York

Springer-Verlag Berlin Heidelberg New York
a member of BertelsmannSpringer Science+Business Media GmbH

http://www.springer.de

© Springer-Verlag Berlin Heidelberg 2002
Printed in Germany

Typesetting: Camera-ready by author, data conversion by Steingräber Satztechnik GmbH, Heidelberg
Printed on acid-free paper   SPIN: 10846482   06/3142   5 4 3 2 1 0

Foreword

Constraint Programming is an approach for modeling and solving combinatorial problems that has proven successful in many applications. It builds on techniques developed in Artificial Intelligence, Logic Programming, and Operations Research. Key techniques are constraint propagation and heuristic search.

Constraint Programming is based on an abstraction that decomposes a problem solver into a reusable constraint engine and a declarative program modeling the problem. The constraint engine implements the required propagation and search algorithms. It can be realized as a library for a general purpose programming language (e.g. C++), as an extension of an existing language (e.g. Prolog), or as a system with its own dedicated language.

The present book is concerned with the architecture and implementation of constraint engines. It presents a new, concurrent architecture that is far superior to the sequential architecture underlying Prolog. The new architecture is based on concurrent search with copying and recomputation rather than sequential search with trailing and backtracking. One advantage of the concurrent approach is that it accommodates any search strategy. Furthermore, it considerably simplifies the implementation of constraint propagation algorithms since it eliminates the need to account for trailing and backtracking.

The book investigates an expressive generalization of the concurrent architecture that accommodates propagation-preserving combinators (known as deep guard combinators) for negation, disjunction, implication, and reification of constraint propagators. Such combinators are beyond the scope of Prolog's technology. In the concurrent approach they can be obtained with a reflective encapsulation primitive.

The concurrent constraint architecture presented in this book has been designed for and realized with the Mozart programming system, where it serves as the basis for new applications and tools. One example presented in this book is the well-known Oz Explorer, a visual and interactive constraint programming tool.

The author of this book, Christian Schulte, is one of the leading experts in constraint technology. He also is one the creators of the Mozart programming system. His book is a must read for everyone seriously interested in constraint technology.

December 2001   Gert Smolka

Preface

Constraint programming has become the method of choice for modeling and solving many types of problems in a wide range of areas: artificial intelligence, databases, combinatorial optimization, and user interfaces, just to name a few. In particular in the area of combinatorial optimization, constraint programming has been applied successfully to planning, resource allocation, scheduling, timetabling, and configuration.

Central to the success of constraint programming has been the emphasis on programming. Programming makes constraint-based modeling expressive as it allows sophisticated control over generation and combination of constraints. Programming makes an essential contribution to the constraint solving abilities as it allows for sophisticated search heuristics.

Todays constraint programming systems support programming for modeling and heuristics. However, they fall short for programming search strategies and constraint combinators. They typically offer a fixed and small set of search strategies. Search cannot be programmed, which prevents users from constructing new search strategies. Search hard-wires depth-first exploration, which prevents even system developers from constructing new search strategies. Combination is exclusively based on reification which itself is incompatible with abstractions obtained by programming and often disables constraint solving when used for combination.

The main contribution of this book is easy to explain: constraint services such as search and combinators are made programmable. This is achieved by devising *computation spaces* as simple abstractions for programming constraint services at a high level. Spaces are seamlessly integrated into a concurrent programming language and make constraint-based computations compatible with concurrency through encapsulation.

State-of-the-art and new search strategies such as visual interactive search and parallel search are covered. Search is rendered expressive and concurrency-compatible by using copying rather than trailing. Search is rendered space and time efficient by using recomputation. Composable combinators, also known as deep-guard combinators, stress the control facilities and concurrency integration of spaces. Composable combinators are applicable to arbitrary abstractions without compromising constraint solving.

The implementation of spaces is presented as an orthogonal extension to the implementation of the underlying programming language. The resulting implementation is shown to be competitive with existing constraint programming systems.

Acknowledgments

First and foremost, I would like to thank Gert Smolka, for his unfailing support, expert advice, and incomparable enthusiasm during my doctoral research. I am extraordinarily grateful to Seif Haridi, who amongst other things accepted to give his expert opinion on the thesis on which this book is based: I do consider Seif as my second thesis adviser. Gert, in particular, introduced me to constraint programming and research in general, and taught me that striving for simplicity is essential and fun. I am still grateful to Sverker Janson for convincing me that the solve combinator is a suitable starting point for a thesis.

I am lucky to have had the great advantage of having been surrounded by knowledgeable and helpful co-workers: Martin Henz, Tobias Müller, Peter Van Roy, Joachim P. Walser, and Jörg Würtz have been inspiring partners in many fruitful discussions on constraint programming; Michael Mehl, Tobias Müller, Konstantin Popov, and Ralf Scheidhauer shared their expertise in implementation of programming languages and constraints with me; Denys Duchier, Seif Haridi, Martin Henz, Michael Mehl, and Gert Smolka have been helpful in many discussions on spaces. I have benefitted very much from the knowledge – and the eagerness to share their knowledge – of Per Brand, Seif Haridi, Sverker Janson, and Johan Montelius on AKL. Konstantin Popov has been invaluable for this work by implementing the first version of the solve combinator.

Thorsten Brunklaus, Raphaël Collet, Martin Henz, Tobias Müller, Gert Smolka, and Peter Van Roy provided helpful and constructive comments on drafts of this work. Their comments have led to fundamental improvements.

April 2001   Christian Schulte

Table of Contents

1. Introduction

This book presents design, application, implementation, and evaluation of simple abstractions that enable programming of standard and new constraint services at a high level. The abstractions proposed are *computation spaces* which are integrated into a concurrent programming language.

1.1 Constraint Programming

Constraint programming has become the method of choice for modeling and solving many types of problems in a wide range of areas: artificial intelligence, databases, combinatorial optimization, and user interfaces, just to name a few.

The success of constraint programming is easy to explain. Constraint programming makes modeling complex problems simple: modeling amounts to naturally stating constraints (representing relations) between variables (representing objects). Integration into a programming language makes modeling expressive. Adapting models is straightforward: models can be changed by adding, removing, and modifying constraints. Constraint programming is open to new algorithms and methods, since it offers the essential glue needed for integration.

Last but not least, the popularity of constraint programming is due to the availability of efficient constraint programming systems. A constraint programming system features two different components: the *constraints* proper and *constraint services*.

Constraints. Constraints are domain specific. They depend on the domain from which the values for the variables are taken. Popular domains for constraint programming are finite domains (the domain is a finite subset of the integers) [140], finite sets [38], trees [24], records [137], and real intervals [98].

Essential for constraints is *constraint propagation*. Constraint propagation excludes values for variables that are in conflict with a constraint. A constraint that connects several variables propagates information between its variables. Variables act as communication channels between several constraints.

C. Schulte: Programming Constraint Services, LNAI 2302, pp. 1–7, 2002.
© Springer-Verlag Berlin Heidelberg 2002

Constraint Services. Constraint services are domain independent. They support the generation, combination, and processing of constraints. Application development amounts to programming with constraints and constraint services.

Powerful generation of constraints according to possibly involved specifications is essential for large and complex problems. The availability of a programming language for this task contributes to the expressiveness of modeling.

Regardless of how many primitive constraints a system offers, combination of constraints into more complex application-specific constraints is a must. This makes means for combining constraints key components of a constraint programming system.

The most important constraint service is search. Typically, constraint propagation on its own is not sufficient to solve a constraint problem by assigning values to variables. Search decomposes problems into simpler problems and thus creates a search tree. It is essential to control shape as well as exploration of the search tree.

A recent survey on research in constraint programming is [146], an introductory book on programming with constraints is [75], and an overview on practical applications of constraint programming is [150].

1.2 Motivation

A cornerstone for the initial success of constraint programming has been the availability of logic programming systems. They successfully integrated constraints and constraint propagation into programming systems that come with built-in search. Most of todays constraint programming systems are constraint logic programming (CLP) systems that evolved from Prolog: CHIP [31, 2], Eclipse [151], clp(FD) [23] and its successor GNU Prolog [29], and SICStus [15], just to name a few. The CLP-approach to search is adopted by cc(FD) [145]. Jaffar and Maher give an overview on CLP in [57].

Search. All these systems have in common that they offer a fixed and small set of search strategies. The strategies covered are typically limited to single, all, and best-solution search. Search cannot be programmed, which prevents users to construct new search strategies. Search hard-wires depth-first exploration, which prevents even system developers to construct new search strategies.

This has several severe consequences. Complex problems call for new search strategies. Research has addressed this need for new strategies. New strategies such as limited discrepancy search (LDS) [48] have been developed and have shown their potential [152, 17]. However, the development of constraint programming systems has not kept pace with the development of search strategies, since search cannot be programmed and is limited. Even well established strategies such as best-first search are out of reach.

Naturally, the lack of high-level programming support for search is an impediment to the development of new strategies and the generalization of existing strategies.

An additional consequence of the fact that controlling search is difficult, is that tools to support the user in search-related development tasks are almost completely missing. Given that search is an essential ingredient in any constraint programming application, the lack of development support is serious.

Combination. The most prominent technique for constraint combination is constraint reification. Reification reflects the validity of a constraint into a 0/1-variable. Constraints can then be combined using 0/1-variables. Typically, reified constraints are combined by Boolean combinators or by generalizations thereof such as the cardinality combinator [143]. Reified constraints are also known as meta-constraints.

Reification as exclusive combination device is problematic, since it dictates an "all or nothing" policy. All constraints subject to combination must be reified. In particular, combining a conjunction of constraints (a common case) requires reification of each conjunct. This results in a dramatic loss of propagation, as reification disables constraint propagation among the conjuncts. Constraints for which the system offers no reified version along with constructions obtained by programming cannot be reified. This renders programming incompatible with reification, resulting in a dramatic loss of expressiveness.

Concurrency. Integration into today's computing environments which are concurrent and distributed is difficult. The backtracking model for search that has been inherited from Prolog is incompatible with concurrency. Most computations including interoperating with the external world *cannot* backtrack.

1.3 Approach

The approach taken in this book is to devise simple abstractions for the programming of constraint services that are concurrency-enabled to start with and overcome the problems discussed in the previous section.

First-Class Computation Spaces. The abstractions are first-class computation spaces and are tightly integrated into a concurrent programming language. Constraint-based computations are delegated to computation spaces.

Computation spaces are promoted to first-class status in the programming language. First-class status of computation spaces enables direct access to constraint-based computations. The direct access allows powerful control of constraint-based computations and by this simplifies programming.

Encapsulation. Computation spaces encapsulate constraint-based computations which are speculative in nature, since failure due to constraint propagation is a regular event. Encapsulation is a must for making constraint programming compatible with concurrency.

Encapsulation is achieved by a tight integration of spaces into the concurrent programming language together with *stability* as powerful control regime. Stability naturally generalizes the notion of entailment. Entailment is known as a powerful control condition in concurrent execution, which has been first identified by Maher [74] and subsequently used by Saraswat for the cc (concurrent constraint programming) framework [118, 117]. Stability has been first conceived by Janson and Haridi in the context of AKL [59, 44, 58].

Oz Light. Computation spaces are integrated into the Oz Light programming language. The essential features of Oz Light that make the integration of spaces possible are computing with partial information through logic variables, implicit synchronization of computations, explicit concurrency, and first-class procedures.

Oz Light is an idealization of the concurrent programming language Oz that concentrates on the features mentioned above. Smolka discusses in [134] the Oz Programming Model (OPM) on which Oz Light is based. OPM extends the concurrent constraint programming (cc) paradigm [118, 117] by explicit concurrency, first-class procedures, and concurrent state.

Search. Spaces are applied to state-of-the-art search engines, such as plain, best-solution, and best-first search. Programming techniques for space-based search are developed and applied to new and highly relevant search engines. One new search engine is the Oz Explorer, a visual and interactive search engine that supports the development of constraint programming applications. Additionally, spaces are applied to parallel search using the computational resources of networked computers.

Copying and Recomputation. In order to be expressive and compatible with concurrency, search is based on copying rather than on trailing. Trailing is the currently dominating approach for implementing search in constraint programming systems. The book establishes the competitiveness of copying by a rigid comparison with trailing.

Recomputation is used as an essential technique for search. Recomputation saves space, possibly at the expense of increased runtime. Recomputation can save runtime, due to an optimistic attitude to search.

The combination of recomputation and copying provides search engines that offer a fundamental improvement over trailing-based search for large problems. The book introduces adaptive recomputation as a promising technique for solving large problems.

Composable Constraint Combinators. Spaces are applied to composable constraint combinators. Composable means that combinators programmed from spaces can combine arbitrary computations, including computations already spawned by combinators. Combinators obtained from spaces are applicable to all statements of the programming language without sacrificing constraint propagation. It is shown how to make composable combinators compatible with reification while avoiding its "all or nothing" approach. Constraint combinators are shown to have a surprisingly simple implementation with spaces. Composable combinators are also known as deep-guard combinators.

Implementation. The book presents an implementation for first-class computation spaces as a conservative extension of an implementation for Oz Light. The implementation is factored into orthogonal support for multiple constraint stores as needed by multiple spaces, stability, space operations, and search. Copying leads to a simple implementation of search.

The implementation model serves as foundation for spaces in the Mozart implementation of Oz [92]. Mozart is a production quality system and is shown to be competitive with existing constraint programming systems.

1.4 Outline

The book consists of five parts. The book's structure and dependencies between chapters are sketched in Figure 1.1.

Setting the Stage. Chapter 2 introduces constraint inference methods and identifies underlying concepts for: constraint propagation, constraint distribution, search, and best-solution search. Chapter 3 introduces Oz Light and relates it to full Oz.

Search. Chapter 4 introduces a simplification of first-class computation spaces for programming search engines. Their design takes the primitives identified in Chapter 2 as input.

Chapters 5 through 7 develop essential techniques for programming search engines from spaces. Plain search engines are introduced in Chapter 5. Best-solution search and some generalizations are discussed in Chapter 6. Different recomputation strategies are developed and evaluated in the following chapter.

The remaining two chapters in this part apply the previously developed techniques to search engines that are new to constraint programming. The Oz Explorer, a visual and interactive constraint programming tool, is discussed in Chapter 8. Parallel search engines that exploit the resources of networked computers are presented in Chapter 9.

Combinators. Chapter 10 presents the full model of computation spaces that enable the programming of composable constraint combinators. The next chapter applies spaces to a wide range of combinators and develops concomitant programming techniques.

Setting the Stage

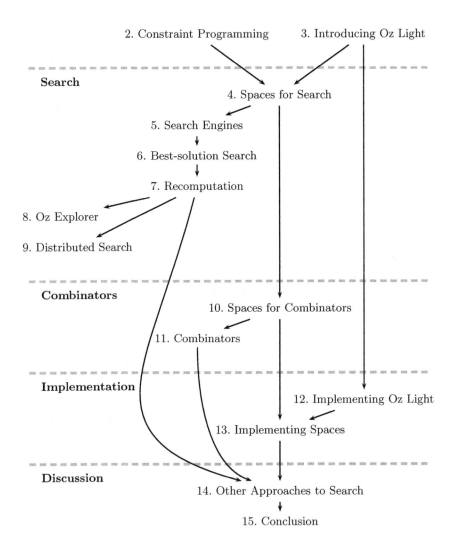

Fig. 1.1. Organization of the book.

Implementation. Chapter 12 lays the foundation for the implementation of computation spaces by outlining an implementation architecture of Oz Light. The next chapter discusses the implementation of first-class computation spaces together with extensions such as support for different constraint domains.

Discussion. The last part is concerned with evaluating and discussing the book's results. This includes comparison with other approaches to search and in particular a detailed comparison with trailing (Chapter 14). Chapter 15 concludes by summarizing the main contributions and presenting concrete ideas for future work.

1.5 Source Material

Part of this book's material has already been published in the following articles:

- Christian Schulte. Parallel Search Made Simple. Techniques for Implementing Constraint Programming Systems, 2000 [126].
- Christian Schulte. Programming Deep Guard Concurrent Constraint Combinators. Practical Aspects of Declarative Languages, 2000 [127].
- Christian Schulte. Comparing Trailing and Copying for Constraint Programming. International Conference on Logic Programming, 1999 [123].
- Christian Schulte. Programming Constraint Inference Engines. International Conference on Principles and Practice of Constraint Programming, 1997 [122].
- Christian Schulte. Oz Explorer: A Visual Constraint Programming Tool. International Conference on Logic Programming, 1997 [121].

Computation spaces build on a previous treatment of the so-called solve combinator, which shares important aspects with spaces. Section 4.7 relates spaces to the solve combinator. The solve combinator has been published in the following articles:

- Christian Schulte and Gert Smolka. Encapsulated Search in Higher-order Concurrent Constraint Programming. International Symposium on Logic Programming, 1994 [128].
- Christian Schulte, Gert Smolka, and Jörg Würtz. Encapsulated Search and Constraint Programming in Oz. Principles and Practice of Constraint Programming, 1994 [129].

The first implementation of the solve combinator has been done by Konstantin Popov as masters thesis [106] under my supervision.

2. Constraint Programming

This chapter introduces essential constraint inference methods and clarifies why constraint programming matters. The inference methods covered are constraint propagation, constraint distribution, and search.

2.1 Constraints

Constraints express relations between variables. Operationally, constraints compute the values for variables that are consistent with the constraints. That is, constraints compute with partial information about values of variables.

Computation Spaces. Computation with constraints takes place in a *computation space*. A computation space consists of propagators (to be explained later) connected to a constraint store. The *constraint store* stores information about values of variables as a conjunction of basic constraints.

propagator $\qquad \cdots \qquad$ propagator

constraint store

Basic Constraints. A *basic constraint* is a logic formula interpreted in some fixed first-order structure. The remainder of this chapter restricts its attention to finite domain constraints. A *finite domain constraint* is of the form $x \in D$ where the *domain* D is a subset of some finite subset of the natural numbers. If D is the singleton set $\{n\}$, the constraint $x \in \{n\}$ is written $x = n$, and x is said to be determined to n. Other domains common in constraint programming are trees and finite sets.

Non-basic Constraints. *Non-basic constraints* typically express relations between several variables and are computationally involved. In order to keep operations on constraints efficient, non-basic constraints are not written to the constraint store. Examples for non-basic finite domain constraints are $x + y \leq z$ or that the values of variables $x_1, \ldots, x_n$ are distinct.

Constraint Propagation. A non-basic constraint is imposed by a *propagator*. A propagator is a concurrent computational agent that amplifies the information in the constraint store by *constraint propagation*. In the context of finite domain constraints, amplification narrows variable domains.

C. Schulte: Programming Constraint Services, LNAI 2302, pp. 9–14, 2002.
© Springer-Verlag Berlin Heidelberg 2002

Suppose a store that contains the constraint ϕ and a propagator that imposes the constraint ψ. The propagator can *tell* (or *propagate*) a basic constraint β to the store, if β is *adequate* ($\phi \wedge \psi$ entails β), *new* (ϕ does not entail β), and *consistent* ($\phi \wedge \beta$ is consistent). Telling β to a store containing ϕ updates the store's constraint to $\phi \wedge \beta$.

Consider the space sketched to the right. The propagator imposing $x > y$ can propagate $x \in \{4, 5\}$ and $y \in \{3, 4\}$. The propagator remains: not all of its information is propagated yet.

$$x > y$$
$$|$$
$$\boxed{x \in \{3, 4, 5\} \wedge y \in \{3, 4, 5\}}$$

A propagator imposing ψ becomes *entailed*, if it detects that ψ is entailed by the constraint store ϕ. It becomes *failed*, if it detects that ψ is inconsistent with ϕ. A propagator that detects entailment disappears. A propagator detects entailment and failure at latest, if all its variables are determined.

A space S is *stable*, if no further constraint propagation in S is possible. A stable space S is *failed*, if S contains a failed propagator. A stable space S is *solved*, if S contains no propagator.

Propagators communicate through the constraint store by shared variables. Suppose that $x + 3 = y$ propagates $x \in \{1, 2, 3\}$ and $y \in \{4, 5, 6\}$. The right propagator then propagates $x \in \{1, 2\}$. Narrowing x triggers the left propagator

$$x + 3 = y \qquad y - 2 \times x > 1$$
$$\searrow \qquad \swarrow$$
$$\boxed{x \in \{1, \ldots, 6\} \wedge y \in \{1, \ldots, 6\}}$$

again to tell $y \in \{4, 5\}$. Now the right propagator is triggered again, telling $x = 1$ which in turn triggers the first propagator to tell $y = 4$. Since both x and y are determined, the propagators disappear.

2.2 Search

Constraint propagation alone is typically not sufficient to solve a constraint problem: a space can become stable, but neither solved nor failed.

$$x \neq y \qquad x \neq z \qquad y \neq z$$
$$\searrow \qquad | \qquad \swarrow$$
$$\boxed{x \in \{1, 2\} \wedge y \in \{1, 2\} \wedge z \in \{1, 2\}}$$

The constraints to the right are unsatisfiable, but no further propagation is possible. Similarly, if the domains for x, y, and z are $\{1, 2, 3\}$, the problem has solutions albeit no further propagation is possible.

Constraint Distribution. To proceed in this situation *distribution* is used: proceed to spaces that are easier to solve, but retain the same set of solutions. Distributing a space S with respect to a basic constraint β creates two spaces: One is obtained by adding the constraint β to S, the other by adding $\neg\beta$ to S. It is crucial to choose β such that both β and $\neg\beta$ trigger further constraint propagation. The constraints β and $\neg\beta$ are called *alternatives*. Distribution is also known as *labelling* or *branching*.

In the context of finite domain constraints, a possible strategy to distribute a space is as follows. Select a variable x with a non-singleton domain

D and a number $n \in D$, and then distribute with $x = n$. This strategy is known as *naive* distribution strategy. A popular refinement is *first-fail*: select a variable with smallest domain.

Search Trees. Search is a complete method for solving finite domain constraint problems. Initially, create a space that contains the basic constraints and propagators of the problem to be solved. Then propagate constraints until the space becomes stable. If the space is failed or solved, search is done. Otherwise, the space is *distributable*.

Search proceeds by distributing the space. Iterating constraint propagation and distribution leads to a tree of spaces, the *search tree*. Each node in the search tree corresponds to a computation space. Leaves correspond to solved or failed spaces. Throughout the book failed spaces are drawn as boxes □, solved spaces as diamonds ◇, and distributable spaces as circles ○.

Exploration. An important property of the setup is that the search tree is defined entirely by the distribution strategy. An orthogonal issue is how to explore the search tree. Possible strategies are depth-first or breadth-first exploration.

A program that implements exploration is called *search engine*. The strategy implemented by the search engine is referred to by *search strategy*. Besides of different strategies engines can offer a great variety of functionality:

– Search for a single solution, several solutions, or all solutions (Chapter 5).
– Interactive and visual search (Chapter 8).
– Search in parallel making use of networked computers (Chapter 9).

Figure 2.1 shows the search tree for the space sketched to the right, where naive distribution with order x, y, and z is used. The figure's right part shows the store of the non-failed nodes after constraint propagation.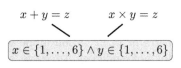

Node	x	y	z
1	$\{1,\ldots,5\}$	$\{1,\ldots,5\}$	$\{2,\ldots,6\}$
3	$\{2,\ldots,5\}$	$\{1,2,3\}$	$\{3,\ldots,6\}$
4	2	2	4
5	$\{3,4,5\}$	$\{1,2\}$	$\{4,5,6\}$

Fig. 2.1. Example search tree.

Best-Solution Search. For a large class of applications it is important to find a best solution with respect to an application-dependent criterion. The naive approach to first compute all solutions and then select the best is infeasible. Typically, the number of solutions grows exponentially with problem size. But even in case the number of solutions remains manageable, one can do better.

The idea of best-solution search is to employ information from an already computed solution to reduce the remaining search space. The information is expressed by constraints: after a solution has been found, the additional constraint that a next solution must be better is taken into account. By this additional constraint, the search tree can become considerably smaller: the constraint *prunes* the search space.

As an example, consider searching for a solution where the value of x is largest. Suppose that there is already a solved space S that assigns x to n. To ensure that search starting from a space S' yields a better solution, the propagator $x > n$ is added to S'. Searching for a solution of S' then can only yield a solution with a value of x that is greater than n. The space S' is *constrained* by *injecting a constraint* to S'.

Branch-and-bound best-solution search works as follows. The search tree is explored until a solution S is found. During exploration, a node is *open*, if it is distributable but has not been distributed yet. All spaces corresponding to open nodes are constrained. This is repeated whenever exploration yields a next solution. If exploration is complete, the solution found last is best.

Consider the space sketched to the right. The goal is to search for a solution with z is largest. Again, naive distribution with order x, y, and z is used. Figure 2.2(a) shows the search tree explored with a left-most depth-first strategy until a first solved space is found. The value for z is 1, Spaces 3 and 4 are constrained by $z > 1$. The figure shows the nodes after injection. Node 5 gets failed ($x \geq z$ propagates $z = 1$) by adding $z > 1$. Adding $z > 1$ to Node 3 propagates $z = 2$.

$$x \geq z \qquad\qquad y > z$$

$$\boxed{x \in \{1,2,3\} \wedge y \in \{1,2,3\} \wedge z \in \{1,2,3\}}$$

Figure 2.2(b) shows the complete search tree. Continuing exploration by distributing Node 3 creates Nodes 6 and 7, of which Node 6 is a new and better solution (z has value 2). The constraint $z > 2$ is added to Node 7 which leads to failure (z is 2). Hence, the best solution is $x = 2 \wedge y = 3 \wedge z = 2$.

The search tree for best-solution search is determined also by the order in which the nodes of the search tree are explored. This is in contrast to "plain" search, where distribution alone determines the search tree. However, this is the very idea of best-solution search: use previous solutions to prune the remaining search space.

In the example above, Nodes 5 and 7 are both pruned. The pruning constraints also interact with the distribution strategy. The strategy possibly considers other constraints for distribution. In the example, this is not the

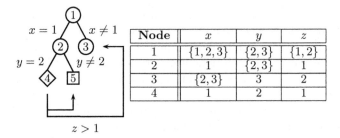

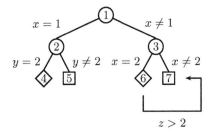

(a) After first solution found.

(b) After second (best) solution found.

Fig. 2.2. Trees for left-most branch-and-bound search.

case. However, the constraint $z > 2$ excludes distribution with respect to z in the subtree issuing from Node 3.

2.3 Programming

The need for programming with constraints arises at the application and at the service level.

Programming Applications. A characteristic task in programming applications with constraints is the creation of constraints according to a problem specification. This process normally extends over several levels of abstraction. It requires programming to compose application-dependent constraints from system-provided constraints with the help of combination mechanisms. Common combination mechanisms are Boolean *combinators* such as disjunction and negation.

The anatomy of a constraint-based application is sketched to the right. The *script* is programmed from constraints and distributors needed for the application. A search engine solves the script.

Programming Services. Services are the abstractions required to program applications, such as propagators, distributors, combinators, and search engines. This book concentrates on *programming* services rather than on *applying* services. More specifically, the interest is on programming generic services such as search and combinators as opposed to domain-specific propagators.

Programming with Spaces. Programming languages allow to build abstractions in a hierarchical fashion, ranging from simple abstractions programmed from primitives to sophisticated abstractions programmed from simpler abstractions. To get the whole process started, the right primitives and their smooth integration into a programming language is essential.

Here, computation spaces have been introduced as central concept for constraint inference. This is also the route of choice in the remainder of the book. The integration of spaces together with primitive operations on spaces in the concurrent programming language Oz Light is described.

The primitives of interest are essentially those required to program constraint services such as search engines and combinators. The exposition of search in Section 2.2 already identified central operations: space creation taking a script as input, control to decide whether a space is solved, failed, or distributable, space distribution, and constraint injection. While in this chapter the intuitive aspects are central, for the purpose of integration into a programming language the concern is to design a small set of abstractions that enables the simple programming of constraint services.

Concurrency. Concurrency plays two important roles in our approach. Firstly, the underlying programming language is concurrent to fit the needs of todays concurrent and distributed computing environments. Secondly, constraint propagation itself is inherently concurrent. Therefore, control of propagation must be concurrency-aware and combination mechanisms must be concurrent.

3. Introducing Oz Light

This chapter introduces Oz Light as the programming language used in the remainder of this book. Oz Light is introduced as language on which the design of computation spaces builds. Extensions and syntactic convenience for programming space-based constraint services are sketched.

3.1 Overview

The essential features of Oz Light are the following.

Partial Information. Oz Light computes with partial information accessible through logic variables. Information on values of variables is provided by constraints.

Implicit Synchronization. Execution implicitly synchronizes until enough information is available on the variables of a statement. Missing information blocks execution. New information resumes execution ("data-flow synchronization").

Explicit Concurrency. Computation is organized into multiple concurrent threads. Threads are created explicitly by the programmer.

First-Class Procedures. Procedures are first-class citizens: they can be passed as arguments and stored in data structures. Procedures maintain reference to external entities by lexical scoping.

This chapter gives a brief overview. A tutorial introduction to Oz is [43]. Smolka discusses in [134] the Oz Programming Model (OPM), on which Oz Light is based. Oz Light extends the concurrent constraint programming (cc) paradigm [117, 118] by explicit concurrency and first-class procedures.

Section 3.2 introduces Oz Light. The following section covers standard concepts (such as exception handling), or concepts orthogonal to the basic setup of Oz Light (such as ports and finite domain constraints). Section 3.4 introduces syntactic convenience to increase the readability of programs used in the remainder of the book. The last section relates Oz Light to full Oz.

It is recommended to read all of Section 3.2 before proceeding with the remaining chapters. The material contained in Section 3.3 is best read as the need arises.

C. Schulte: Programming Constraint Services, LNAI 2302, pp. 15–27, 2002.

3.2 Oz Light: Basics

Computation in Oz Light takes place in a *computation space*. A computation space features multiple *threads* computing over a shared *store*. A thread is the control struc-

ture of a sequential computation. The store contains the data structures with which the threads compute. Variables connect threads to the data structures in the store. The so-far single space is called *toplevel space*.

A thread is a stack of statements. A thread reduces by trying to reduce its topmost statement. Reduction automatically synchronizes until the store contains sufficient information on the variables of the topmost statement.

3.2.1 The Store

The store has two compartments: the *constraint store* and the *procedure store*. The constraint store contains logic formulas that represent information about values of variables. The procedure store contains procedures which are created as computation proceeds.

Procedures are data structures but not values. They are connected to the constraint store by primitive values called names. The procedure store maps names to procedures.

The constraint store contains information about values of variables represented by a conjunction of *basic constraints*. Basic constraints are logic formulas interpreted in a fixed first-order structure, called the *universe*. The elements of the universe are the values with which threads compute. Variables are ranged over by x, y, and z.

The Universe. The universe contains *integers*, *atoms*, *names*, and *rational trees* [24] constructed from *tuples* of values. Values are ranged over by v and integers by i.

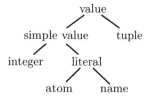

Names (ranged over by ξ and η) are primitive entities that have no structure. There are two special names **true** and **false** that represent the respective truth values.

Atoms are symbolic values that have identity as defined by a sequence of characters. Examples for atoms are ´atom´, ´nil´, and ´|´. A *literal* is either a name or an atom. Literals are ranged over by l. A *simple value* is either a literal or an integer. Simple values are ranged over by s.

A *tuple* $l(v_1 \ldots v_n)$ consists of a single *label* l (a literal) and *fields* $v_1, \ldots, v_n$ with $n > 0$. The number of fields n is called the tuple's *width*.

Lists are constructed from tuples and atoms as follows. A list is either the empty list (the atom **nil**) or a pair of an element (the head) and a list (the tail). A pair is a binary tuple ´|´$(x\ y)$ which can be written infix as $x|y$.

Constraints. A constraint ϕ is a conjunction of basic constraints β. The constraint store contains a constraint which defines the values the variables can take. A basic constraint β is one of the following:

- $x = s$, which is interpreted that the value of x is the simple value s.
- $x = l(y_1 \ldots y_n)$, which is interpreted that x is a tree with label l and subtrees defined by y_1 through y_n.
- $x = y$, which is interpreted that the values of x and y are the same.

In the following a constraint store is often identified with its constraint.

Satisfiability and Entailment. A constraint ϕ is *satisfiable*, if $\exists\, \phi$ is valid in the universe. A constraint ϕ *entails* a constraint ψ, if $\phi \to \psi$ is valid in the universe. A constraint ϕ *disentails* a constraint ψ, if ϕ entails $\neg\psi$.

Determined, Aliased, and Constrained Variables. A variable x is *determined* by a constraint ϕ, if there exists a simple value s such that ϕ entails $x = s$, or if there exists a literal l and a natural number $n > 1$ such that ϕ entails $\exists \overline{y}\; x = l(y_1 \ldots y_n)$. In the former case, x is determined to $s/0$, in the latter, to l/n. A variable x is *aliased* to a variable y by a constraint ϕ, if $x \neq y$ and ϕ entails $x = y$. A variable x is aliased, if there exists a variable y to which x is aliased. A variable x is *constrained* by a constraint ϕ, if x is determined or aliased by ϕ.

If the constraint is omitted by which variables are determined, aliased, or constrained, the constraint stored by the constraint store is assumed.

Telling Constraints. *Telling* a basic constraint β to a constraint store ϕ updates the constraint store to contain $\phi \wedge \beta$, provided that $\phi \wedge \beta$ is satisfiable. This means that it is only possible to tell basic constraints that leave the store satisfiable. Starting out from an empty store (that is $\top$), the constraint store maintains the invariant to be satisfiable. In case an attempt to tell a basic constraint would render the store unsatisfiable, the attempt is said to be unsuccessful.

Dependency. A variable x *depends on a variable* y with respect to a constraint ϕ, $x \triangleleft_\phi y$, if either y is aliased to x by ϕ, or ϕ entails $\exists \overline{z}_1 \overline{z}_2\; x = l(\overline{z}_1 z \overline{z}_2)$ and $z \triangleleft_\phi y$. Analogously, a variable x *depends on a name* ξ with respect to a constraint ϕ, $x \triangleleft_\phi \xi$, if there exists a variable y that is determined to ξ by ϕ and $x \triangleleft_\phi y$.

3.2.2 Threads

A thread is a stack of statements. A thread can only reduce if its topmost statement can reduce. Reduction of the topmost statement pops the statement and can also:

- Tell information to the constraint store.
- Create a new procedure and enter it into the procedure store.

– Push statements on the stack.
– Create a new thread.

Statements are partitioned into *synchronized* and *unsynchronized* statements. Reduction of an unsynchronized statement takes place independently of the information in the constraint store. In contrast, synchronized statements can only reduce if the constraint store provides sufficient information.

Information in the constraint store is accessed by variables: a statement *synchronizes* or *suspends* on variables. A thread itself synchronizes or suspends, if its topmost statement synchronizes. The set of variables a statement σ and its thread T synchronizes on, is called its *suspension set* and is denoted by $\mathcal{S}(\sigma)$ and $\mathcal{S}(T)$.

Reduction of threads is fair. If a thread can reduce because either its topmost statement is unsynchronized or otherwise the constraint store contains sufficient information, it eventually will reduce.

If the last statement of a thread reduces and pushes no new statement, the thread *terminates* and ceases to exist. If the topmost statement of a thread can reduce, the thread is *runnable*. Otherwise the thread is *suspended*. A suspended thread becomes runnable by *waking* or by *resuming*. A runnable thread becomes suspended by *suspending* the thread.

The *current thread* is the thread whose topmost statement is being reduced. By pushing a statement σ, it is meant that σ is pushed on the current thread.

3.2.3 Statements

The core statements of Oz Light are shown in Figure 3.1. Their reduction is as follows.

Empty Statement. The empty statement

$$\texttt{skip}$$

reduces without any effect and is unsynchronized.

Tell. A tell statement

$$x = v$$

is unsynchronized. Its reduction attempts to tell $x = v$ to the constraint store. An unsuccessful attempt raises an exception, which is discussed later.

Sequential Composition. A sequential composition statement

$$\sigma_1 \ \sigma_2$$

is unsynchronized. It reduces by pushing σ_2 and then σ_1.

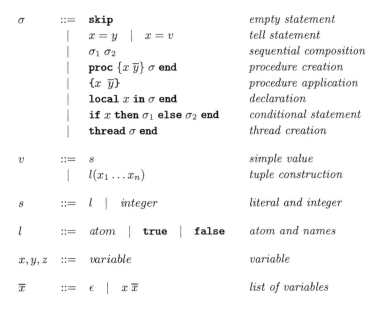

Fig. 3.1. Statements of Oz Light.

Declaration. A declaration statement

$$\textbf{local } x \textbf{ in } \sigma \textbf{ end}$$

is unsynchronized. It creates a fresh variable y and reduces by pushing $\sigma[y/x]$, where x is replaced by y in $\sigma[y/x]$.

Procedure Creation. A procedure creation statement

$$\textbf{proc } \{x \, \overline{y}\} \, \sigma \textbf{ end}$$

is unsynchronized. Its reduction chooses a fresh name ξ, stores the procedure $\lambda\overline{y}.\sigma$ under the name ξ in the procedure store, and pushes $x = \xi$.

The statement σ is the *body* and the variables $\overline{y}$ are the *(formal) arguments* of the procedure. The arguments $\overline{y}$ are required to be linear, that is, no variable occurs twice in $\overline{y}$. The variables that occur free in σ but not in $\overline{y}$ are the procedure's *free variables*.

The notation $\xi \mapsto \lambda\overline{y}.\sigma$ is used for a procedure $\lambda\overline{y}.\sigma$ stored under a name ξ. Since ξ is fresh, storing the procedure under the name ξ maintains the invariant that the procedure store is a mapping of names to procedures.

Procedure Application. A procedure application statement

$$\{x \, \overline{y}\}$$

synchronizes on the variable x. Reduction requires x to be determined to a name ξ with $\xi \mapsto \lambda\overline{z}.\sigma$ and the number of actual parameters $\overline{y}$ must match the number of formal parameters $\bar{z}$. Reduction pushes $\sigma[\bar{y}/\bar{z}]$ where the formal parameters are replaced by the actual parameters.

Conditional. A conditional statement

$$\text{if } x \text{ then } \sigma_1 \text{ else } \sigma_2 \text{ end}$$

synchronizes on the variable x. If x is determined to **true**, reduction proceeds by pushing σ_1. Otherwise, reduction proceeds by pushing σ_2.

Thread Creation. A thread creation statement

$$\textbf{thread } \sigma \textbf{ end}$$

is unsynchronized. Its reduction creates a new thread that consists of the statement σ.

3.3 Oz Light Continued

This section is concerned with additional features of Oz Light. The statements that are discussed in this section are listed in Figure 3.2.

3.3.1 Primitive Operations

Equality Test. The equality test

$$x = (y == z)$$

synchronizes until $y = z$ is either entailed or disentailed. If $y = z$ is entailed, reduction proceeds by pushing $x = \textbf{true}$. If $y = z$ is disentailed, reduction proceeds by pushing $x = \textbf{false}$.

Determination Test. The test whether a variable is determined

$$\{\texttt{IsDet } x \ y\}$$

is unsynchronized. If x is determined, $y = \textbf{true}$, otherwise $y = \textbf{false}$ is pushed.

Indeterminate Synchronization. The operation

$$\{\texttt{WaitOr } x \ y\}$$

synchronizes until x or y is determined. Its reduction has no effect.

Arithmetic Operations. Unary minus ($\tilde{}$) is an example for arithmetic operations.

$$x = \tilde{}y$$

Its reduction synchronizes until y is determined to an integer i. Reduction proceeds by pushing $x = -i$.

$$\sigma \quad ::= \quad x = (y == z) \qquad\qquad\qquad \textit{equality test}$$

σ	::=	$x = (y == z)$	*equality test*
	|	{IsDet x y}	*determination test*
	|	{WaitOr x y}	*indeterminate synchronization*
	|	$x = {}^\sim y \quad\|\quad x = y\ (\ +\ \|\ -\ \|\ >\ )\ z$	*arithmetic operations*
	|	{Width x y} | {Label x y}	*tuple operations*

(a) Primitive operations.

σ	::=	**try** σ_1 **catch** x **then** σ_2 **end**	*try statement*
	|	**raise** x **end**	*raise statement*

(b) Exception handling.

σ	::=	{NewPort x y}	*port creation*
	|	{Send x y}	*message sending*

(c) Ports.

σ	::=	$x::y$	*domain tell*
	|	{FdReflect x y}	*domain reflection*

(d) Finite domain constraints.

Fig. 3.2. Statements of Oz Light, continued.

Tuple Operations. The operation

$$\{\text{Width } x\ y\}$$

synchronizes until x is determined to s/n. Reduction proceeds by pushing $y = n$. Similarly, {Label x y} proceeds by pushing $y = s$.

A common abstraction is {Wait x} that synchronizes on x being determined. This can be expressed by either {WaitOr x x} or by

```
proc {Wait X}
    if X==1 then skip else skip end
end
```

A convenient abstraction First that does indeterminate synchronization on two variables is programmed as follows. {First X Y Z} blocks until at least one of X and Y becomes determined. If Z is **true** (**false**), X (Y) is determined.

```
proc {First X Y Z}
    {WaitOr X Y} {IsDet X Z}
end
```

Example 3.1 (Indeterminism). WaitOr adds indeterminism to Oz Light. In the following example, it is indeterminate whether Z is determined to 1 or 2:

```
thread X=1 end thread Y=1 end
if {First X Y} then Z=1 else Z=2 end
```

3.3.2 Exceptions

Try Statement. A try statement

$$\textbf{try } \sigma_1 \textbf{ catch } x \textbf{ then } \sigma_2 \textbf{ end}$$

is unsynchronized. It first pushes **catch** x **then** σ_2 **end** and then σ_1.

catch x **then** σ **end** is used to define the semantics of exceptions. A programmer is not allowed to use this statement in programs.

Catch Statement. A catch statement

$$\textbf{catch } x \textbf{ then } \sigma \textbf{ end}$$

is unsynchronized. Its reduction has no effect.

Raise Statement. A raise statement

$$\textbf{raise } x \textbf{ end}$$

is unsynchronized. All statements until **catch** y **then** σ **end** (included) are popped. Then $\sigma[x/y]$ is pushed.

3.3.3 Ports and Active Services

Ports provide message sending for communicating concurrent computations. A port maintains an ordered stream of messages ("mailbox"). A **Send**-operation on the port appends a message to the end of the stream. The stream of messages then can be incrementally processed as new messages arrive.

Ports are accommodated like procedures: variables refer to names, which refer to ports. For that matter, the store is extended by a *port store* as third compartment.

Port Creation. The statement

$$\{\texttt{NewPort } x \ y\}$$

is unsynchronized. Its reduction creates a fresh name ξ and stores $[x]$ in the port store under the name ξ. Reduction proceeds by pushing $y = \xi$. The variable x stored by the port $[x]$ is the tail of the message stream.

As with procedures, $\xi \mapsto [x]$ refers to a port with stream x stored under name ξ.

Message Sending. The statement

$$\{\text{Send } x \ y\}$$

is synchronized. Reduction requires x to be determined to $\xi \mapsto [z_1]$. The message y is added to the stream as follows. A new variable z_2 is created and the port is updated to $\xi \mapsto [z_2]$. Reduction proceeds by pushing $z_1 = y|z_2$. Reduction maintains the invariant that the variable stored in the port is the tail of the message stream.

Ports have been initially conceived in the context of AKL [60]. They have been adopted in Oz, but as abstractions obtained from cells and not as primitives [134]. Cells are a more primitive concept to capture concurrent state. Here ports rather than cells are made primitive, since ports are extended as the presentation of the book proceeds in a way that is not captured easily by the cell-based construction.

Note that ports are an additional source of indeterminism. If messages to a port are sent by multiple concurrent threads, the order of messages on the stream is indeterminate.

Active Services. The **Send**-operation on ports can be easily extended to deal with replies to messages. Rather than sending the message, a pair of message and answer is sent. The answer is a logic variable which serves as place holder for the answer.

This idea is captured by the following procedure definition:

```
proc {SendRecv P X Y}
    local M in M=´#´(X Y) {Send P M} end
end
```

A common abstraction for communicating concurrent computations is the use of *active services*. An active service is hosted by a thread of its own. It processes messages that arrive on a stream and computes answers to the messages.

The procedure **NewService** as shown in Figure 3.3 takes a procedure P and computes a new procedure **ServiceP** that encapsulates message sending. All messages are served in a newly created thread by **Serve**. A more readable version using syntactic convenience is available in Figure 3.4.

Active services combine concurrency control with latency tolerance. All messages are served sequentially which makes concurrency control simple. Message sending is asynchronous and the service's client can immediately continue its computation. Only when needed, possibly much later, the client automatically synchronizes on the answer.

3.3.4 Finite Domain Constraints

For finite domain constraints, the constraint store supports the basic constraint $x \in D$. Here $D \subseteq \{0, \ldots, \hat{n}\}$, where $\hat{n}$ is a sufficiently large natural number. The constraint is interpreted that the value of x is an element of D.

```
proc {Serve XYs P}
   if XYs==nil then skip else
      local XY XYr X Y in
         XYs=XY|XYr XY=´#´(X Y) {P X Y} {Serve XYr P}
      end
   end
end
proc {NewService P ServiceP}
   local XYs Po in
      {NewPort XYs Po}
      thread {Serve XYs P} end
      proc {ServiceP X Y}
         {SendRecv Po X Y}
      end
   end
end
```

Fig. 3.3. Creating active services.

A variable x is *kinded* by a constraint ϕ, if x is not determined by ϕ and ϕ entails $x \in \{0, \ldots, \hat{n}\}$. Accordingly, a variable x is *constrained*, if x is determined, kinded, or aliased. Note that a variable can be both kinded and aliased.

Telling Domains. The statement

$$x::y$$

synchronizes until y is determined to a list $n_1 | \cdots | n_k | \mathtt{nil}$ of natural numbers. It reduces by attempting to tell the basic constraint $x \in \{n_1, \ldots, n_k\}$.

Domain Reflection. The statement

$$\{\mathtt{FdReflect}\ x\ y\}$$

synchronizes until x is kinded or determined. Suppose that $D = \{n_1, \ldots, n_k\}$ is the smallest set for which $x \in D$ is entailed and that $n_1 < n_2, \ldots, n_{k-1} < n_k$. Reduction proceeds by pushing a statement that constructs an ordered list containing $n_1, \ldots, n_k$:

```
local z₁ local z′₁ ··· local zₖ local z′ₖ in
    y = z₁|z′₁  z₁ = n₁  ···  z′ₖ₋₁ = zₖ|z′ₖ  zₖ = nₖ  z′ₖ =nil
end ··· end
```

Propagators. For the purposes of this book it is sufficient to regard a propagator as a thread that implements constraint propagation. More information on the integration of propagators into Oz can be found in [84, 157].

3.4 Syntactic Convenience

This section introduces syntactical convenience to ease programming of constraint services in the remainder of the book. A tutorial account on Oz syntax is [43], a rigid treatment is [51].

Declaration. Multiple variables can be introduced simultaneously:

> `local X Y in` σ `end` $\Rightarrow$ `local X in local Y in` σ `end end`

If a declaration statement comprises the body of a procedure definition or the branch of a conditional, `local` and `end` can be omitted:

> `proc {P} Y in` σ `end` $\Rightarrow$ `proc {P} local Y in` σ `end end`

Declaration can be combined with initialization through tell statements:

> `local X=5 in` σ `end` $\Rightarrow$ `local X in X=5` σ `end`

Functional Notation. The statement $z = \{x\ \overline{y}\}$ abbreviates $\{x\ \overline{y}\ z\}$. Motivated by this abbreviation, $\{x\ \overline{y}\}$ is said to return z. Similarly, nesting of tuple construction and procedure application avoids declaration of auxiliary variables. For example:

> ```
>                          local Y Z in
> X=b({F N+1})      ⇒          Y=N+1 X=b(Z) {F Y Z}
>                          end
> ```

Tuple construction is given precedence over procedure application to allow more procedure definitions to be tail recursive. The construction is extended analogously to other statements, allowing statements as expressions. For example:

> ```
> X=local Y=2 in
>       {P Y}          ⇒     local Y=2 in X={P Y} end
>    end
> ```

Procedure definitions as expressions are tagged with a dollar sign ($) to distinguish them from definitions in statement position:

> ```
> X=proc {$ Y}
>       Y=1            ⇒     proc {X Y} Y=1 end
>    end
> ```

Procedure definitions can use functional notation by using **fun** rather than **proc**, where the body of a functional definition is an expression:

> `fun {Inc X} X+1 end` $\Rightarrow$ `proc {Inc X Y} Y=X+1 end`

Lists. Complete lists can be written by enclosing the elements in square brackets: [1 2] abbreviates 1|2|nil and ´|´(1 ´|´(2 nil)).

Infix Pairs. The label ´#´ for pairs ´#´(X Y) can be written infix: X#Y.

Pattern Matching. Programming with tuples and lists is greatly simplified by pattern matching. A pattern matching conditional

$$\textbf{case } x \textbf{ of } l(y_1 \ldots y_n) \textbf{ then } \sigma_1 \textbf{ else } \sigma_2 \textbf{ end}$$

is an abbreviation for

```
if {Width x}|{Label x} == n|l then
    y₁...yₙ in x = l(y₁...yₙ) σ₁
else σ₂
end
```

The else part is optional and defaults to **else skip**. Multiple clauses are handled sequentially, for example:

```
case X                          case X of f(Y) then σ₁
of f(Y) then σ₁        ⇒        else case X of g(Z) then σ₂ end
[] g(Z) then σ₂                 end
end
```

try-statements are also subject to pattern matching. For example:

```
                                try σ₁ catch Y then
try σ₁                              case Y of f(X) then σ₂
catch f(X) then σ₂     ⇒            else raise Y end
end                                 end
                                end
```

Figure 3.4 shows as an example a version of active services that is considerably easier to read and understand than the formulation shown in Figure 3.3.

3.5 Relation to Full Oz

The presentation of Oz Light is targeted at the actual need of the book. The most prominent features of full Oz missing in the previous exposition are as follows.

Values: Records. Full Oz offers a richer universe and contains in particular floating point numbers and records. Records generalize tuples in that subtrees can be referred to by name rather than by position only. Information on the universe can be found in [133]. More on records can be found in [137].

Finite Set Constraints. In addition to finite domain constraints, full Oz offers constraints ranging over finite sets of integers [82, 85].

```
fun {SendRecv P X}
   Y in {Send P X#Y} Y
end

proc {Serve XYs P}
   case XYs of (X#Y)|XYr then
      Y={P X} {Serve XYr P}
   end
end
fun {NewService P}
   XYs Po={NewPort XYs}
in
   thread {Serve XYs P} end
   fun {$ X}
      {SendRecv Po X}
   end
end
```

Fig. 3.4. Active services using syntactic convenience.

Futures. Full Oz offers futures as read only variants of logic variables [78]. Futures provide reliability for programming abstractions such as active services and support demand-driven execution.

Cells. Concurrent state is provided by cells in full Oz. Cells implement mutable bindings of names to variables. The development of the Oz Programming Model by Smolka [134] covers cells and clarifies their relation to ports.

Classes and Objects. Full Oz supports concurrent objects that are obtained by instantiation from classes [49]. Classes are subject to multiple inheritances. Objects offer support mutual exclusion by monitors. By this they are an alternative to active services for structuring concurrent computations.

Distribution. Full Oz supports distributed execution across several computers connected through the Internet. Distribution is discussed in Section 9.2.

Modules. Full Oz offers a powerful module system that supports separate compilation and both static and dynamic linking [35]. The module system serves also as access control mechanism for distributed execution, which is discussed in Section 9.2.

4. Spaces for Search

This chapter introduces a simplified model of first-class computation spaces for programming search engines.

4.1 Overview

Computation spaces have been introduced as central mechanism for search in Chapter 2. This chapter follows the idea and integrates spaces into Oz Light to program search engines. It presents a simplified model for spaces that is sufficient for search engines. Chapter 10 generalizes spaces to more expressive constraint services.

The integration of spaces into Oz Light is concerned with three major issues.

Language Integration. Spaces are integrated smoothly into Oz Light in order to ease programming. Ease of programming is facilitated by promoting spaces to first-class citizens in the programming language. Search engines are then programmed from operations on first-class computation spaces.

Encapsulation. Constraint-based computations are *speculative* in that failure is a regular event. Speculative computations need encapsulation in a concurrent context. Using backtracking for speculative computations as in Prolog is unfeasible. Most computations including interoperating with the external world cannot backtrack.

Operations. The ease of programming search engines depends on which operations on spaces are available. This makes the design of suitable operations crucial.

The introduction of computation spaces is organized as follows:

Local Computation Spaces. Speculative constraint-based computations are delegated to local computation spaces. Their setup is discussed in Section 4.2.

First-Class Spaces. First-class spaces provide a handle to encapsulated speculative computations and operations for creation, modification, and access (Section 4.3).

C. Schulte: Programming Constraint Services, LNAI 2302, pp. 29–44, 2002.
© Springer-Verlag Berlin Heidelberg 2002

Control and Status. Computation spaces employ stability as simple control
condition. Stability and stability-based control operations are introduced
in Section 4.4.

Search. Operations for distribution and search are introduced in Section 4.5.

Communication. Section 4.6 refines active services to support communication
across space boundaries while obeying encapsulation.

Section 4.7 discusses a previous approach for programming search. Section 4.8 provides a brief summary of computation spaces for programming
search engines.

4.2 Local Computation Spaces

The key idea for encapsulation is to delegate the execution of a speculative computation to a local computation space. A
local computation space features, like the
toplevel space, local variables, local names,
and a private constraint store. Execution in
a local space resembles execution in the toplevel.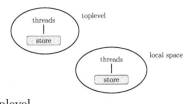

Each space S provides the same constituents as the toplevel: threads,
store, local variables, and local names. Each entity e (thread, variable, name,
and procedure) is *situated* in exactly one space S, its *home* (space) $\mathcal{H}(e)$.
The home of the current thread is referred to as *current space*. Similarly, the
notion *current store* is used. Notions such as determined, aliased, and kinded
that are with respect to a constraint refer by default to the current store.

The basic idea of local spaces is that computations in a local space perform
as they do at the toplevel. However, some points need conservative extension.

Freshness and Visibility. The set of variables and names for each space are
disjoint. This means that a fresh variable or a fresh name is fresh with respect
to all spaces.

In a local space, variables and names of the toplevel and of the local space
itself are visible. Visible means that computations can access them.

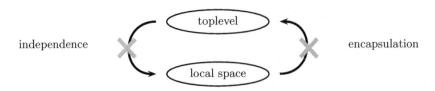

Fig. 4.1. Independence and encapsulation for local spaces.

Independence. The setup of spaces in this chapter makes two simplifications. Firstly, no nested spaces are allowed: spaces cannot be created inside spaces. Secondly, after creation, the space becomes *independent* of the toplevel (Figure 4.1). Independence is guaranteed by the fact that only determined toplevel variables are visible in a local space, non-determined variables are ruled out. This invariant is satisfied by space creation and is discussed later.

Procedure Application. When a procedure application reduces in a local space, the appropriate procedure is taken from the union of the local procedure store and the toplevel procedure store. As a consequence of the disjointness of names, the procedure to be applied is uniquely determined.

Tell. Execution of a tell statement

$$x = v$$

tells $x = v$ in the current space.

Failure. An unsuccessful attempt to tell $x = v$ *fails* a local computation space. Failing the local space stops all computations: all threads in the local space are discarded.

Input and Output. The toplevel is the only space that is designated to execute non-speculative computations. For this reason input and output is allowed in the toplevel only. In all other spaces an attempt to perform input or output raises an exception.

4.3 Space Manipulation

This section is concerned with operations that create new spaces, merge spaces with the toplevel space, and inject computations into existing spaces.

Computation Spaces Are First-Class Citizens. To enable programming, computation spaces are promoted to first-class status: each space S is uniquely referred to by a name ξ (similar to procedures). A space S with first-class reference ξ is accordingly written as $\xi \mapsto S$. Programs can refer to the space S by a variable that is determined to ξ.

4.3.1 Space Creation

A new space is created by

$$\{\texttt{NewSpace}\ x\ y\}$$

Reduction blocks until x satisfies an independence condition that is explained later. A new name ξ is created with the toplevel as home. A new space S is created as follows:

- The *root variable* of S is initialized with a fresh variable z with home S. The root variable serves as entry point to the constraints of S.
- The set of local variables is initialized to contain z. The set of local names and the procedure store are initialized as being empty.
- The constraint store is initialized with the constraints of the toplevel.
- A thread is created in S to execute $\{x\ z\}$.

Finally, the statement $y = \xi$ is pushed.

The procedure passed to NewSpace is called *script* and defines which computation is performed. To speculatively execute a statement σ, a new space is created by:

```
S={NewSpace proc {$ _} σ end}
```

Example 4.1 makes use of the root variable.

Independence Condition. An essential simplification in this chapter is that a local space has no access to not yet determined variables of the toplevel space. This restriction ensures that computations in a local space after their initial setup are independent of the toplevel.

This is achieved by restricting scripts to not refer to undetermined variables via the free variables of the script. The restriction is not formalized further, since the full model in Chapter 10 does not impose this purely didactic restriction.

Synchronizing on Spaces. Operations on spaces other than NewSpace need to synchronize on a variable x being determined to a name ξ that refers to the space S. Execution is said to synchronize until x is $\xi \mapsto S$.

4.3.2 Merging Spaces

Access to a speculative computation combines two aspects. Firstly, access to the result of a speculative computation via the root variable. Secondly, access to the entire speculative computation itself by removing the space serving as encapsulation barrier.

The following primitive combines both aspects

$$\{\text{Merge } x\ y\}$$

Synchronizes on x being $\xi \mapsto S$. If S is failed, an exception is raised. Otherwise, S is merged with the toplevel space as follows:

- S is marked as merged.
- The set of local variables of the toplevel is updated to include the local variables of S. The same happens with the local names.
- Similarly, the procedure store of the toplevel is updated to include the mappings of S's procedure store.
- $y = z$ is pushed, where z is the root variable of S.
- All constraints of S are told in the toplevel.

Example 4.1 (Speculative Execution). The function **F** (to be read as unary procedure) can be speculatively evaluated by

    S={NewSpace F}

To access the result by **X**, the space **S** is merged:

    X={Merge S}

This is not yet convincing! Before accessing the result by merging the space, the space's status must be checked: Has the speculative computation failed? Has it terminated successfully? These issues are dealt with in Section 4.4.

4.3.3 Injecting into Spaces

It can become necessary to spawn new computations in an already existing space. As an example consider best-solution search as discussed in Section 2.2: A space gets a constraint "injected" that it must yield a better result than the previous solution.

This is captured by the primitive

$$\{\text{Inject } x \ y\}$$

that synchronizes on: x is $\xi \mapsto S$ and y refers to a procedure which satisfies the same independence condition as discussed for space creation. If S is failed, the operation does nothing. Otherwise, a new thread in S is created that executes $\{y \ z\}$ where z is the root variable of S.

Example 4.2 (Killing a Space). A particular application of **Inject** is to kill a speculative computation

    proc {Kill S}
       {Inject S proc {$ _} fail end}
    end

by injecting **fail** into S. **fail** abbreviates a statement that raises failure, for example **local** X **in** X=1 X=2 **end**.

4.4 Control and Status

Example 4.1 shows that it is essential to know when and if a speculative computation has reached a stable state.

Stability. The definition of when a space has reached a stable state is straightforward. A space is *runnable*, if it contains a runnable thread. A space is *stable*, if it is not runnable. According to this definition, a failed space is stable. A space is *succeeded*, if it is stable and not failed.

Status Access. Controlling a space S requires an operation that blocks until S becomes stable and then returns its status.

$$\{\texttt{Ask } x \ y\}$$

Reduction synchronizes on x being $\xi \mapsto S$ and S being stable. It reduces according to S's status: if S is failed (merged, succeeded), $y = \texttt{failed}$ ($y = \texttt{merged}$, $y = \texttt{succeeded}$) is pushed. Section 4.5.2 extends Ask to accommodate for distributor threads as needed for search.

Example 4.3 (Example 4.1 Reconsidered). With Ask it is possible to program a satisfactory abstraction that speculatively evaluates an expression. The following procedure takes again a unary procedure P that returns the computation's result.

```
fun {Speculate P}
   S={NewSpace P}
in
   if {Ask S}==failed then nil else [{Merge S}] end
end
```

Speculate returns the empty list (nil) in case the speculative computation has been unsuccessful. Otherwise a singleton list containing the result is returned.

Ask synchronizes on stability of a space and then returns its status. Section 10.4.2 presents a simpler and more expressive design that does not require synchronization on spaces but reuses synchronization on variables.

4.5 Search

To support search, spaces need operations for distribution and exploration. An important goal is to make distribution programmable and to decompose it into orthogonal primitives. Distribution is generalized as follows:

Arbitrary Statements. Distribution is not limited to be with respect to a single constraint. Instead an arbitrary number of statements, called *alternatives*, are allowed.

Explicit Cloning. Distribution is programmed from cloning spaces and committing a space to a particular alternative.

Problem-Independent Exploration. Alternatives are problem-dependent and thus require access to the constraints inside a space. For exploration, it is sufficient to select alternatives by number. The number-based selection protocol makes the search engine orthogonal to the script to be solved.

Factoring script and exploration is an important design principle. It follows in spirit one of the main motivations of logic programming which is often referred to by the slogan "algorithm = logic + control", due to Kowalski [69].

4.5.1 Alternatives

A straightforward approach is to use a choice statement for specifying the alternatives with which a space is to be distributed:

choice σ_1 [] $\cdots$ [] σ_n **end**

where the statements $\sigma_1, \ldots, \sigma_n$ define the alternatives. This approach would statically fix the number of alternatives.

The primitive Choose allows an arbitrary number of alternatives:

$$\{\texttt{Choose } x \ y\}$$

Its reduction blocks until x is determined to a natural number n. If $n \leq 0$ an exception is raised. Otherwise, the current thread T is marked as *distributor thread* with n alternatives. If S already contains a distributor thread, an exception is raised. This construction ensures that there can be at most one distributor for a space.

The variable y will be determined to a number between 1 and n. The determination is controlled by a different primitive Commit that is used to program exploration of alternatives. The primitive is discussed in Section 4.5.4.

For convenience,

choice σ_1 [] $\cdots$ [] σ_n **end**

abbreviates

```
case {Choose n}
of 1 then σ₁
[] ···
[] n then σₙ
end
```

Multiple Distributors are Considered Harmful. A different and seemingly more expressive design would be to allow multiple distributors per space. This design alternative has been explored in different flavors in earlier implementations of Oz and has been identified to be a common source of hard to find programming errors.

The first flavor is to leave the order of distributors undefined. This renders search unpredictable: explored nodes and search tree size depend on which distributor is considered first. This in particular collides with recomputation (Chapter 7) which presupposes that exploration can be redone deterministically.

The second flavor is to order distributors. A dynamic order that records distributors in order of creation does not offer any improvement in a concurrent setting. A static order can improve this, but is difficult to define: all concurrent events must be ordered, including injection into spaces. In addition, a static order is costly to implement [58, 89]. On the other hand, expressiveness is still unsatisfactory, see Section 11.6.

As a consequence, a simple but expressive way to employ multiple distributors is to explicitly program the order. Multiple distributors are then executed by a single thread in a well defined order. Since the base language is indeterministic (due to determination test and message sending), indeterministic creation of distributors is still possible. The point is that with at most one distributor, this type of error is considerably less likely.

4.5.2 Distributable Spaces

A space is *distributable*, if it is stable and contains a distributor thread T. A distributable space is said to have n alternatives, if its distributor thread has n alternatives. Consequently, a stable space is succeeded, if it is neither failed nor distributable.

The additional stable state is taken into account by Ask for status access as follows.

$$\{\text{Ask } x \ y\}$$

If x refers to a distributable space which has n alternatives, reduction pushes $y = \text{alternatives}(n)$.

4.5.3 Synchronizing on Stability

Typically, a distributor creates alternatives that reflect the current information available in the constraint store. For example, a distributor following a first-fail strategy gives preference to a variable with smallest domain. Which variable to select is decided best after all constraint propagation is done, that is, after the current space is stable.

This requires a primitive that allows a thread to synchronize on stability of the current space. As a followup to the discussion above, it is useful to restrict the number of threads that can synchronize on stability to at most one. Along the same lines, a space can have either a distributor thread or a thread that waits for stability.

Therefore Choose is extended such that it offers synchronization for stability in case the number of alternatives is one. If a space S becomes distributable and has a single alternative, reduction immediately proceeds as follows: the distributor thread T contains as its topmost statement {Choose x y} and x is determined to 1. This statement is replaced by pushing $y = 1$. This possibly makes both T and S runnable again. A space that is distributable and has a single alternative is called *semi-stable*. Note that a space that becomes semi-stable, directly becomes runnable again by reduction of the Choose statement.

In the following the procedure WaitStable is used to synchronize on stability that is programmed from Choose:

```
proc {WaitStable}
   {Choose 1 _}
end
```

Example 4.4 (Programming Distribution). `Distributor` takes a list of finite domain variables to be distributed:

```
proc {Distributor Xs}
   {WaitStable}
   case {SelectVar Xs} of [X] then N={SelectVal X} in
      choice X=N [] X\=:N end {Distributor Xs}
   else skip
   end
end
```

`WaitStable` is employed to synchronize the shaded statement on semi-stability. This ensures that variable and value selection take place after constraint propagation.

After synchronizing on stability, `SelectVar` selects a variable `X` that has more than one value left, whereas `SelectVal` selects a possible value `N` for `X`. The binary choice states that either `X` is equal to `N` or different from `N`. The first-fail strategy, for example, implements `SelectVar` as to return a variable with smallest domain and `SelectVal` as to return the smallest value.

4.5.4 Committing to Alternatives

For exploration, a space must be reduced with alternatives defined by `Choose`. This is done with:

$$\{\text{Commit } x \ y\}$$

Its reduction synchronizes on x being $\xi \mapsto S$ and S being stable and not merged. Additionally, it synchronizes on $y = n$ for some natural number n. An exception is raised, if S is not distributable, if n is less than one, or if n is greater than the number of alternatives of S.

Otherwise, the distributor of S contains a statement $\{\text{Choose } z \ z'\}$. This statement is replaced by pushing $y = z'$.

At first sight, it seems not essential that `Commit` synchronizes on stability. Typically, before `Commit` is applied, `Ask` has been used to test that the space is indeed distributable. There are search engines (recomputation being one particular example, Chapter 7) that repeatedly apply `Commit`. For these engines synchronization of `Commit` on stability is convenient and excludes a great deal of programming errors by design.

4.5.5 Cloning Spaces

A space is cloned by

$$\{\text{Clone } x \ y\}$$

Its reduction synchronizes on x being $\xi \mapsto S$ and S being stable. It reduces by creating a clone $\xi' \mapsto S'$ of S with ξ' being a fresh name. Variables and names in S are consistently renamed to fresh variables and fresh names. Reduction pushes $y = \xi'$.

Stability is essential for cloning. It fights combinatorial explosion by ensuring that all computation is done once and for all before cloning. As will become clear in Section 13.5.2, stability is essential for an efficient implementation of cloning.

Example 4.5 (Clone and Merge). Cloning performs a consistent renaming of local variables and names. As a consequence, the statement

```
C={Clone S} in {Merge S}={Merge C}
```

can possibly raise failure! As an example for a space S to exhibit this behavior consider

```
S={NewSpace proc {$ P} proc {P} skip end end}
```

Example 4.6 (Distribution). Suppose that S refers to a distributable space with two alternatives. Then S is distributed by

```
fun {Distribute S}
   C={Clone S} in {Commit S 1} {Commit C 2} [S C]
end
```

where S is the space obtained by distribution with the first alternative and C the space obtained by distribution with the second alternative.

From distribution it is only a small step to provide a first blueprint of a search engine programmed from spaces.

Example 4.7 (All-Solution Exploration). Suppose that S refers to a space that has been created for a script to be solved by search. Then all-solution exploration that takes S as input and returns a list of all succeeded spaces representing solutions is as follows:

```
fun {Explore S}
   case {Ask S}
   of failed then nil
   [] succeeded then [S]
   [] alternatives(2) then [S1 S2]={Distribute S} in
      {Append {Explore S1} {Explore S2}}
   end
end
```

Here Append concatenates two lists. Note that Explore is restricted to distributable spaces with two alternatives.

Example 4.8 (Partial Evaluation with Clone). Cloning can be seen as partial evaluation: the result stored in a stable space can be reused as many times as required. In particular, local variables are automatically renamed, whenever the space is cloned.

The following procedure sketches this idea. It takes a script P as input and returns a procedure that on application returns what P would return on application, provided that encapsulated execution of P becomes stable:

```
fun {Evaluate P}
   S={NewSpace P}
in
   if {Ask S}==failed then proc {$ _} fail end
   else proc {$ X} {Merge {Clone S} X} end
   end
end
```

4.5.6 Refining Commit

Commit selects a single alternative. In later chapters, in particular in Chapters 5 and 6, it will become apparent that it is useful to be able to select alternatives at a finer granularity. Rather than committing to a single alternative, it is beneficial to discard some alternatives (or by abusing language, to commit to a number of alternatives).

To this end, Commit is refined as follows:

$$\{\text{Commit2 } x \; y_1 \; y_2\}$$

which synchronizes on x being $\xi \mapsto S$ and S being stable. Additionally, it synchronizes on y_1 and y_2 being determined to natural numbers n_1 and n_2. If $1 \leq n_1 \leq n_2 \leq m$ does not hold, where m is the number of alternatives of S, an exception is raised. The idea is that only the alternatives $n_1, \ldots, n_2$ remain, while the numbering observed by Choose is maintained. If $n_1 = n_2$, reduction coincides with that of Commit.

Otherwise, suppose the distributor thread contains {Choose z_1 z_2} as its first statement. Then this statement is replaced by

```
local X Y in X=n₂ − n₁ + 1 {Choose X Y} z₂=Y+n₁ − 1 end
```

Here, X refers to the new number of alternatives, whereas z_2 is obtained by adding an appropriate offset.

Rather than using Commit2 directly, the following convenient definition of Commit is employed:

```
proc {Commit S X}
   case X of N1#N2 then {Commit2 S N1 N2}
   else {Commit2 S X X}
   end
end
```

4.6 Situated Procedure Calls: Services Reconsidered

The setup disallows communication between local computation spaces and the toplevel space. Even the full model for spaces that is discussed in Chapter 10 will restrict communication across spaces such that it is compatible with encapsulation. For some applications this setup is too strict. Consider the following situations:

Best-First Search. Best-first search associates a cost value with each node of the search tree. The cost value must be computed within the space, since it normally depends on the script's variables and constraints. The value itself must be available to the search engine and must be communicated to the toplevel space.

Database Access. A script might require access to external data. As an example, think of products and their associated costs from a warehouse database. The script cannot access the database directly. The request must be sent to a service in the toplevel space. The answer must be communicated back to the script's space. This scenario has remarkable similarity with remote procedure calls (RPC) used in distributed computing: the computation (the service) is stationary, while the arguments and the result of the call are transmitted across the network (across spaces). This justifies to refer to this technique as *situated procedure call* (SPC).

Mapping the two example situations to ports and services, they just correspond to the operations `Send` and `SendRecv` on ports as introduced in Section 3.3.3.

In both situations, the idea to clone and merge the space to get access is infeasible. The space of interest is typically not stable and thus cannot be cloned. Moreover, in situations where cloning would be applicable, it is inappropriate. It is far to costly to clone the entire space to access a cost value or a small message.

Sending Messages Across Spaces. The message x to be sent to a port that is situated at the toplevel must be sendable. Intuitively, x is sendable to S, if x does not refer to variables and names which are local to S.

A variable x is *sendable* from S with store ϕ to the toplevel space, if there is no variable y with $x \triangleleft_\phi y$ and y is not determined, and there is no name ξ with $x \triangleleft_\phi \xi$ and ξ is situated in S ($\triangleleft$ is introduced in Section 3.2.1).

In case x is sendable from S to the toplevel space, all constraints on x must be made available to a variable x' that is situated in the toplevel. The constraints are made available by cloning them from S to the toplevel. As becomes clear in Section 13.7, the implementation of sending comes for free in that cloning is not needed.

A send statement

$$\{\texttt{Send } x \ y\}$$

reduces as follows. It synchronizes on x being $\xi \mapsto [z]$. If $\mathcal{H}(\xi)$ is the current space, reduction proceeds as described in Section 3.3.3.

An exception is raised if y is not sendable from the current space to the toplevel space. Otherwise, y is sent to x. The message is then appended to the port's stream as follows. The port store is updated to $\xi \mapsto [z']$ and reduction proceeds by injecting $z = y|z'$ into the toplevel space. Since the port store is updated immediately, sequential order of messages is guaranteed.

Getting Answers. The procedure `SendRecv` shown in Section 3.3.3 that returns an answer is programmed from `Send`. This is not longer possible in the context of sending across spaces. If the sender is situated in a local space and the port is situated in the toplevel, the answer is computed in the toplevel and hence the variable to refer to the answer must be situated in the toplevel as well.

Therefore, `SendRecv` becomes a primitive operation. Its definition is straightforward and follows the idea that the variable to take the answer is situated in the port's home.

4.7 Previous Work: Solve Combinator

Previous work by me and Smolka introduced the *solve combinator* [128, 129, 133]. The solve combinator spawns a local computation space and resolves choices by returning them as procedures. Computation spaces subsume the solve combinator and avoid its severe limitations.

```
fun {Solve P}
   S={NewSpace P}
in
   case {Ask S}
   of failed then failed
   [] succeeded then
      solved(proc {$ X} {Merge {Clone S} X} end)
   [] alternatives(N) then C={Clone S} in
      distributed(proc {$ X} {Commit S 1}    {Merge S X} end
                  proc {$ X} {Commit C 2#N} {Merge C X} end
                  if N==2 then last else more end)
   end
end
```

Fig. 4.2. The solve combinator programmed from computation spaces.

The solve combinator programmed from spaces is shown in Figure 4.2. It takes a script, creates a space that executes the script, and returns information that depends on the space status. It combines the abstractions of Example 4.3 for speculative execution, of Example 4.8 for providing a solution, and of Example 4.6 for distribution.

Spaces provide a more expressive and natural abstraction for programming constraint services. The main disadvantage of the solve combinator is that it hardwires distribution. This prevents all but the most straightforward services: while most but not all services discussed in Chapters 5 and 6 can be programmed from the solve combinator, the services in Chapters 7 to 11 are out of reach.

Example 4.9 (All-Solution Exploration Reconsidered). The solve combinator can be regarded as a convenient abstraction to program *simple* search engines. The following example shows all-solution exploration as presented in Example 4.7 programmed with the solve combinator.

```
fun {ExploreAll P}
   case {Solve P} of failed then nil
   [] solved(P) then [P]
   [] distributed(P1 P2 _) then
      {Append {ExploreAll P1} {ExploreAll P2}}
   end
end
```

ExploreAll returns a list of unary procedures rather than a list of computation spaces and, again, is limited to binary alternatives only. The main difference to be observed is that distribution is fully automatic with the solve combinator: this accounts for both its elegance and its lack of expressiveness.

4.8 Summary

This section summarizes operations on first-class computation spaces as introduced in this chapter.

NewSpace : *Script → Space*
> Creates a new space with a thread executing the script applied to the root variable (Section 4.3.1).

Inject : *Space × Script*
> Injects a thread executing the script applied to the root variable (Section 4.3.3).

Merge : *Space → Any*
> Merges a space with the toplevel and returns the root variable (Section 4.3.2).

Ask : *Space → Status*
> Synchronizes until a space becomes stable and then returns the status (Section 4.4). Figure 4.3(a) summarizes the states of a computation space.

Clone : *Space → Space*
> Creates a clone of a stable space (Section 4.5.5).

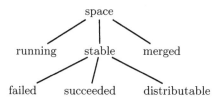

(a) Relation between states.

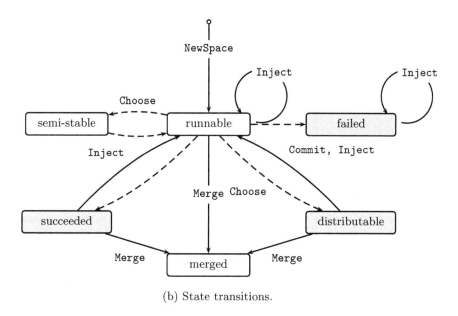

(b) State transitions.

Fig. 4.3. Summary of space states and transitions.

`Commit` : *Space* × *Alternative*
> Commits a distributable space to alternatives of its distributor (Section 4.5.4).

Figure 4.3(b) summarizes the transitions between states of a computation space. States that correspond to stable spaces are shaded. Transitions performed by operations applied to first-class spaces are depicted with solid lines. Transitions performed by computation situated in the space are depicted with dashed lines. The picture shows that stable states are indeed stable with respect to computation inside the space: no dashed edge starts from a stable state.

Relation to Mozart. The operations on computation spaces as discussed here are fully implemented in Mozart (Version 1.2). They are available in a module

`Space` that is loaded on demand at runtime. To try any of the examples presented in this book, it is sufficient to write `Space.ask` rather than `Ask`, for example.

5. Search Engines

This chapter presents simple state-of-the-art search engines. The chapter's focus is on familiarizing the reader with basic techniques for programming search engines.

5.1 Depth-First Search

The most basic search strategy is depth-first search (DFS): explore the search tree left-most depth-first until a first solution is found. In the following, discussion is limited to distributors with two alternatives, the general case is discussed in Section 5.3.

Exploration. The procedure DFE (as abbreviation for depth-first exploration) is shown in Figure 5.1. DFE takes a space as argument and tries to solve it following a depth-first strategy. The procedure is similar to that shown in Example 4.7 and is discussed here again to show how to control exploration until the first solution is found.

```
fun {DFE S}
   case {Ask S}
   of failed then nil
   [] succeeded then [S]
   [] alternatives(2) then C={Clone S} in
      {Commit S 1}
      case {DFE S} of nil then {Commit C 2} {DFE C}
      [] [T] then [T]
      end
   end
end
```

Fig. 5.1. Depth-first one-solution exploration.

If no solution is found, but search terminates, the empty list is returned. Otherwise, a singleton list with the succeeded computation space is returned. If S is distributable, exploration continues with the first alternative. If this does not yield a solution, a clone is distributed with the second alternative and is solved recursively.

C. Schulte: Programming Constraint Services, LNAI 2302, pp. 45–54, 2002.

The Engine. The procedure DFE is turned into a complete search engine DFS that can be used without any knowledge about spaces as follows:

```
fun {DFS P}
   case {DFE {NewSpace P}} of nil then nil
   [] [S] then [{Merge S}]
   end
end
```

DFS takes a script as input, creates a new space to execute the script, and applies DFE to the newly created space. In case DFE returns a list containing a succeeded space, its root variable is returned as singleton list.

Typically, search engines are not programmed from scratch. The Mozart implementation of Oz offers a search library programmed from spaces [33].

All-Solution Search. The search engine can be adapted easily to all-solution search as in Example 4.7. It is sufficient to replace the shaded lines in Figure 5.1 with:

```
{Commit C 2} {Append {DFE S} {DFE C}}
```

Example 5.1 (Send Most Money). As an example, consider a variation of a popular cryptoarithmetic puzzle: Find distinct digits for the variables S, E, N, D, M, O, T, Y such that $S \neq 0$, $M \neq 0$ (no leading zeros), and $SEND + MOST = MONEY$ holds. The puzzle's script is shown in Figure 5.2.

Execution of Root ::: 0#9 tells the basic constraints that each element of Root is an integer between 0 and 9. The propagator FD.distinct enforces all list elements to be distinct, whereas the propagators S\=:0 and M\=:0 enforce the variables S and M to be distinct from 0. The variables for the letters are distributed (by FD.distribute) according to a first-fail strategy.

Applying the search engine DFS to Money returns [[9 3 4 2 1 0 5 7]].

```
proc {Money Root}
   S E N D M O T Y
in
   Root = [S E N D M O T Y] Root ::: 0#9
   {FD.distinct Root}
   S\=:0 M\=:0
                 S*1000 + E*100 + N*10 + D
               + M*1000 + O*100 + S*10 + T
   =: M*10000 + O*1000 + N*100 + E*10 + Y
   {FD.distribute ff Root}
end
```

Fig. 5.2. A program for the $SEND + MOST = MONEY$ puzzle.

5.2 Simplifying Control: Exceptions

Depth-first search for a single solution has a simple termination condition: either exploration is complete, or a solution is found. The procedure DFE in Figure 5.1 keeps on testing the latter condition. This leads to a nesting of conditional statements. A simpler approach is to replace testing by raising an exception in case a solution is found (Figure 5.3). The exception contains the solution found.

```
proc {DFE S}
   case {Ask S}
   of failed then skip
   [] succeeded then raise [S] end
   [] alternatives(2) then C={Clone S} in
      {Commit S 1} {DFE S} {Commit C 2} {DFE C}
   end
end
```

(a) Exploration.

```
fun {DFS P}
   try {DFE {NewSpace P}} nil
   catch [S] then [{Merge S}]
   end
end
```

(b) Engine.

Fig. 5.3. Depth-first search engine using exceptions.

The benefits of using exceptions become even more apparent for engines that consist of more than a single procedure. With testing, each individual procedure must test whether to continue exploration. Examples that in particular benefit from exceptions are limited discrepancy search (Section 5.6) and best-first search (Section 5.7).

5.3 Binarization

The procedure DFE as shown in Figure 5.3 handles two alternatives only. A straightforward way to deal with an arbitrary number of alternatives is by an additional procedure NE (Figure 5.4). To use NE, the procedure DFE must be adopted as follows:

```
[] alternatives(N) then {NE S 1 N}
```

```
proc {NE S I N}
   if I==N then
      {Commit S N} {DFE S}
   else C={Clone S} in
      {Commit C I} {DFE C} {NE S I+1 N}
   end
end
```

Fig. 5.4. Exploring alternatives from 1 to n.

A simpler way is to use *binarization* by splitting alternatives. Different binarization strategies are sketched in Table 5.1. Binarization trades an additional procedure similar to NE for additional commit-operations (c in the table). Section 13.8 provides evidence that commit-operations are efficient. Since two alternatives is the most common case, binarization is a simple and gracefully degrading technique.

Table 5.1. Binarization of n-ary distributors ($n > 2$).

Strategy	Operations	c	f
None	Figure 5.4	n	1
Left	{Commit S 1#(N-1)} {Commit C N}	$2n-2$	$n-1$
Balanced	M=N **div** 2 **in** {Commit S 1#M} {Commit C (M+1)#N}	$2n-2$	$\lfloor \log_2 n \rfloor$
Right	{Commit S 1} {Commit C 2#N}	$2n-2$	1

c Number of commit-operations.
f Number of commit- (clone)-operations for *first* alternative.

Taking runtime into account, balanced binarization looks most promising followed by right binarization. For single-solution search, right binarization has the additional advantage that only one commit- and clone-operation are needed for the first alternative. This is good when search for the first solution almost never goes wrong.

Memory consumption yields an even more compelling argument for right binarization. Using f clone-operations to compute the first alternative also implies that f spaces must be kept in memory during exploration of the first alternative. Therefore right binarization is preferable. In Section 6.3 it is argued that right binarization has further advantages for branch-and-bound best-solution search.

To incorporate right binarization into the search engine for depth-first search, it is sufficient to replace the shaded part in Figure 5.3 by:

```
[] alternatives(N) then C={Clone S} in
   {Commit S 1} {DFE S} {Commit C 2#N} {DFE C}
```

5.4 Multiple Solutions

Section 5.1 sketches how to program all-solution search. In general, searching for all solutions is unfeasible. It is more realistic to search for a limited number of solutions.

An additional disadvantage of the all-solution engine sketched in Section 5.1 is that the engine returns solutions only after the entire search tree has been explored. In a concurrent setting, it is natural to output solutions as early as possible such that other threads can start consuming them.

This idea can be combined with using exceptions for termination. As soon as a solution is found, the engine applies a procedure `Collect` to it. `Collect` then controls how to output the solution and checks whether to continue exploration. If all desired solutions have been found, `Collect` raises an exception that terminates the engine.

A simple example for `Collect` that implements single-solution search as before is:

```
proc {Collect S}
   raise [S] end
end
```

A more interesting example is `SearchSome` that searches for a given number n of solutions where the solutions should be available to other threads immediately. When the search engine starts, it immediately returns a logic variable `Xs`. The variable `Xs` refers to the list of solutions. The definition of `Collect` remembers the tail of the list. If a next solution is found, it is appended to the list of solutions. If all n solutions are found, `Collect` raises an exception to terminate exploration.

Demand-driven search is obtained similarly. After a solution is found, the new tail of the list is chosen as a by-need future. Only when a next solution is requested by synchronizing on the future search continues. By-need futures are available in full Oz, although they are not discussed here in more detail (Section 3.5 gives a glance at futures).

5.5 Explicit State Representation

`DFS` shown in Figure 5.3 maintains its state implicitly as statements on a thread. Engines to be discussed later, for example, best-first search in Section 5.7, and in particular parallel search in Chapter 9, require access to spaces that comprise the engine's state.

```
proc {DFE Ss}
   case Ss of nil then skip
   [] S|Sr then
      case {Ask S} of failed then {DFE Sr}
      [] succeeded then raise [S] end
      [] alternatives(2) then C={Clone S} in
         {Commit S 1} {Commit C 2} {DFE S|C|Sr}
      end
   end
end
```

Fig. 5.5. Depth-first search engine with explicit state.

An alternative formulation of DFE that explicitly maintains spaces is shown in Figure 5.5. The engine maintains the state as a stack of spaces (implemented as list). Exploration is performed more eagerly than exploration by the engine in Section 5.1. The reason is that the commit-operation shaded gray is immediately applied after cloning.

A straightforward solution to arrive at the same number of exploration steps is to not store spaces directly. Instead a data structure is used from which the space is computed if desired. A convenient data structure is of course a function that returns the space upon application. The data structure that is suited best depends on the engine.

5.6 Limited Discrepancy Search

Typically, distribution strategies follow a heuristic that has been carefully designed to suggest most often "good" alternatives leading to a solution. This is taken into account by limited discrepancy search (LDS), introduced by Harvey and Ginsberg [48]. LDS has been successfully applied to scheduling [17, 25] and frequency allocation [152].

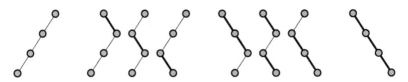

Fig. 5.6. Probes with 0, 1, 2, and 3 discrepancies.

Exploring against the heuristic is called a *discrepancy*. In the setting here, a discrepancy thus amounts to first commit to the second alternative, rather than to the first. LDS explores the search tree with no allowed discrepancy first, then allowing 1, 2, ... discrepancies until a solution is found, or a given limit for the discrepancies is reached. Exploration with a fixed number of allowed discrepancies is called *probing*.

Additionally, LDS makes a discrepancy first at the root of the search tree. This takes into account that it is more likely for a heuristic to make a wrong decision near the root of the tree where only little information is available. If no solution is found, discrepancies are made further down in the tree. Figure 5.6 sketches how LDS probes, where discrepancies are shown by thick vertices (the illustration is adapted from [48]).

```
proc {Probe S M}
   case {Ask S} of failed then skip
   [] succeeded then raise [S] end
   [] alternatives(N) then
      if M>0 then C={Clone S} in
         {Commit S 2#N} {Probe S M-1}
         {Commit C 1}   {Probe C M}
      else
         {Commit S 1} {Probe S 0}
      end
   end
end
```

Fig. 5.7. Probing for LDS.

Figure 5.7 shows **Probe** that implements probing. It takes a space S and the number of allowed discrepancies M as input, and raises an exception being a singleton list containing a succeeded space, if a solution is found. If S is distributable and no more discrepancies are allowed (that is, M is zero) probing continues after committing to the first alternative. Otherwise, a discrepancy is made by committing to the remaining alternatives and probing continues with one allowed discrepancy less. If this does not yield a solution, probing continues by making the discrepancy further down in the search tree. Note that **Probe** uses binarization: the first alternative corresponds to 0 discrepancies, the second alternative to 1 discrepancy, and the i-th alternative to $(i-1)$-discrepancies.

```
proc {Iterate S N M}
   if N==M then {Probe S N}
   else {Probe {Clone S} N} {Iterate S N+1 M}
   end
end
fun {LDS P M}
   try {Iterate {NewSpace S} 0 M} nil
   catch [S] then [{Merge S}]
   end
end
```

Fig. 5.8. Iteration-engine for LDS.

A complete implementation of LDS is obtained straightforwardly from Probe (Figure 5.8). First, a space S running the script P is created. Then application of Probe to a clone of S and the number of allowed discrepancies is iterated until either a solution is found or the discrepancy limit is reached.

It is interesting that LDS is close in structure and shorter (due to exceptions) than the original pseudo-code for probing in [48]. This demonstrates that spaces provide an adequate level of abstraction for search engines of this kind. Results of recent research that has explored improvements of LDS such as ILDS (improved LDS) [68] and variants of LDS such as DDS (depth-bounded discrepancy search) and IDFS (interleaved depth-first search) [153, 80, 81] can be adapted to space-based engines easily.

To iteratively apply exploration is a common technique. The presumably best known example is iterative deepening [66, 67].

5.7 Best-First Search

Distribution makes a *local* heuristic decision based only on the variables of a single space. In some cases it can be preferable to make a *global* decision instead. Best-first search makes global decisions: each node of the search tree has a cost value associated. Exploration always continues with the best node, that is, with the cheapest node.

A best-first search engine takes a cost function in addition to the script. The cost function is problem-specific. Typically, the cost function needs access to the root variable of a script. This is also the most interesting point to be discussed here: the cost is computed inside the space, but must be made available to the search engine that is executed in the toplevel space. With other words, best-first search requires communication across space boundaries. For communication across space boundaries services are used as discussed in Section 4.6.

```
fun {GetCost S F}
   N SF={NewService fun {$ X} N=X unit end}
in
   {Inject S proc {$ R} _={SF {F R}} end} N
end
```

Fig. 5.9. Computing a cost for a space.

Figure 5.9 shows GetCost that takes a space S and a cost function F and returns the cost. It first creates a trivial service SF by application of NewService (Section 3.3.3) that makes the argument of the first invocation of SF available via N. Cost computation is by injecting a thread into S that computes the cost (by application of F to the root variable of S) and applies the service SF to that cost.

```
proc {Insert PQ S F}
   case {Ask S} of failed then skip
   [] succeeded then raise [S] end
   [] alternatives(2) then {Put PQ {GetCost S F} S}
   end
end
```

(a) Insertion according to cost.

```
proc {BFE PQ}
   if {Not {IsEmpty PQ}} then S={Get PQ} C={Clone S} in
      {Commit S 1} {Insert PQ S F}
      {Commit C 2} {Insert PQ C F}
      {BFE PQ}
   end
end
```

(b) Exploration.

Fig. 5.10. Best-first exploration.

The rest of the best-first search engine is straightforward. The engine organizes nodes according to cost using a priority queue. Central parts of best-first exploration are shown in Figure 5.10. Figure 5.10(a) shows how a given space S is inserted into a priority queue PQ (Put enqueues an element into a priority queue) according to S's cost. Note that only distributable nodes are inserted. Failed spaces are ignored and succeeded spaces raise an exception to return a solution. Figure 5.10(b) shows best-first exploration, where F again refers to a cost function. IsEmpty tests whether a priority queue is empty, whereas Get removes and returns the cheapest element.

Example 5.2 (Applying Best-First Search). As an example, best-first search is applied to the $SEND + MOST = MONEY$-problem (Example 5.1). An example cost function is SizeSum: the sum of the sizes of the variables for the letters. The size of a variable is the cardinality of its domain. BFS is invoked as follows:

{BFS SMM SizeSum}

The cost function has a similarity with first-fail distribution: it chooses the space for exploration for which propagation has led to the tightest domain.

Best-first search differs essentially from depth-first exploration. Depth-first exploration allows for a backtracking implementation. Best-first exploration can continue at arbitrary nodes in the search tree. This issue is discussed in more detail in Section 14.2.

A severe drawback of best-first search is that it requires exponential memory in the depth of the search tree for the worst case (similar to breadth-first search). This can be addressed by recomputation strategies (Chapter 7).

Best-first search is just one particular instance of an *informed* search strategy. The point to discuss best-first search in the context of computation spaces is to show how to apply services as technique for "informedness". Other informed search engines are A*-search [47] and its derivatives such as IDA* and SMA*. These strategies are discussed by most textbooks on Artificial Intelligence, for example [97, 114].

6. Best-Solution Search

Best-solution search determines a best solution with respect to a problem-dependent order among solutions. The art of best-solution search is to *prune the search space* as much as possible by previously found solutions. This chapter presents basic techniques and generalizations for best-solution search.

6.1 Constraining Spaces

Essential for best-solution search is to inject into a space an additional constraint that the next solution must be better than all previous solutions. This constraint prunes the search space to be explored for finding a better solution.

The following function takes a binary order procedure O and returns a procedure Constrain.

```
fun {NewConstrain O}
   proc {Constrain S BS}
      OR={Merge {Clone BS}}
   in
      {Inject S proc {$ NR} {O OR NR} end}
   end
in
   Constrain
end
```

Constrain takes a space S and a space BS (the best solution so far). It injects into S that it must yield a better solution than BS. This is implemented by the order O on the constraints accessible from the root variables of the previous solution and S itself.

The solution's constraints are made accessible by merging a clone of BS rather than merging BS itself. This allows to possibly return BS as best solution. Constrain can straightforwardly be optimized by memorizing the solution obtained by merging.

C. Schulte: Programming Constraint Services, LNAI 2302, pp. 55–58, 2002.
© Springer-Verlag Berlin Heidelberg 2002

6.2 Iterative Best-Solution Search

A simple engine for best-solution search is *iterative best-solution search* (IBSS). After a solution is found, search restarts from the original problem together with the constraint to yield a better solution.

Iteration is used as in limited discrepancy search (see Section 5.6). Any single-solution search engine can be used for IBSS. Iteration continues until the search engine does not yield a solution. The best solution (if any) is the solution found last.

6.3 Branch-and-Bound Best-Solution Search

IBSS performs well, if it is easy to find a first solution. If finding a first solution already involves a great deal of search, IBSS is bound to repeat the search in each iteration. In this situation, branch-and-bound search (BAB) can do better, since it avoids repetition.

```
fun {BABE S BS}
   case {Ask S} of failed then BS
   [] succeeded then S
   [] alternatives(N) then C={Clone S} in
      {Commit S 1} {Commit C 2#N}
      local NBS={BABE S BS} in
         if NBS\=BS then {Constrain C NBS} end
         {BABE C NBS}
      end
   end
end
```

Fig. 6.1. Branch-and-bound best-solution search engine.

The procedure BABE (see Figure 6.1) implements exploration for BAB. It takes the space S to be explored and the space BS as the best solution so far. It returns the space for the best solution or nil, if no solution exists. Initially, SolS is nil. The procedure maintains the invariant that S can only lead to a solution that is better than BS. In case S is failed, the so-far best solution is returned. In case S is succeeded, it is returned as new and better solution (which is guaranteed by the invariant).

The central part is shaded: if following the first alternative returns a better solution (the invariant ensures that a different space is better), the space for the second alternative is constrained to yield an even better solution than BS. Note that here the unique identity of spaces and that nil is different from any space is exploited. The latter ensures that Constrain never gets applied to nil.

Binarization (see Section 5.3) has advantages over individually exploring each alternative for BAB. Application of Constrain can potentially prune several alternatives simultaneously rather than prune each alternative individually.

A search engine BABS is obtained easily: it creates a space running the script to be solved, creates a procedure Constrain depending on the order, applies BABE, and possibly returns the best solution.

Example 6.1 (Send Most Money (Example 5.1) Reconsidered). To search for a solution of *SEND + MOST = MONEY* with the most money, that is, *MONEY* is as large as possible, a binary procedure More is defined as follows. It takes two root variables O and N and imposes the constraint that N is better than O.

{BABS Money More} returns [[9 7 8 2 1 0 4 6]] as best solution.

6.4 An Alternative Formulation of BAB

Later chapters present search engines that require explicit access to the search engine's spaces. For this reason and for additional insight, this section presents a formulation of BAB that maintains spaces explicitly.

The BAB engine shown in the previous section uses the identity of spaces to determine whether a space must be constrained. Here, the spaces to be explored are organized on two stacks: the *foreground stack* (f-stack) and the *background stack* (b-stack). Spaces on the f-stack are guaranteed to yield a better solution. Spaces that are not known to guarantee this invariant are on the b-stack.

The engine can be characterized by how it maintains the invariants for the two stacks:

- Initially, the b-stack is empty and the f-stack contains the root space.
- If the f-stack is empty and the b-stack contains S, S is moved to the f-stack after constraining S.
- If a better solution is found, all elements of the f-stack are moved to the b-stack.
- If a space taken from the f-stack is comitted or cloned, it is eligible to go on the f-stack itself.

Taking these facts together yields the program shown in Figure 6.2. The procedure BABE takes the f-stack (Fs), the b-stack (Bs), and the currently best solution (BS).

6.5 Prune-Search: Generalizing BAB

BAB uses the currently best solution to prune the remaining search space. This section shows how to generalize this idea: accumulate information on all

```
fun {BABE Fs Bs BS}
   case Fs of nil then
      case Bs of nil then BS
      [] B|Br then {Constrain B BS} {BABE [B] Br BS}
      end
   [] F|FR then
      case {Ask F} of failed then {BABE Fr Bs BS}
      [] succeeded then {BABE nil {Append Fr Bs} F}
      [] alternatives(N) then C={Clone F} in
         {Commit F 1} {Commit C 2#N} {BABE F|C|Fr Bs BS}
      end
   end
end
```

Fig. 6.2. BABE with explicit state representation.

solutions found so far to prune the rest of the search space. This technique is called *prune-search* (PS).

One particular instance of PS is of course BAB. Accumulation is rather pathological: the information is just the last solution found. Pruning is achieved by injecting the constraint that a solution has to be better than the currently best one.

A different example for PS is searching for all solutions to *SEND + MOST = MONEY* with different amounts of *MONEY*. A naive approach is to search for all solutions and then remove solutions with the same values for *MONEY*. For larger problems, where one is interested in "essentially different" solutions this approach is unfeasible. The accumulated information are the different values for *MONEY*. Initially, the list is empty. Each solution found contributes a new value to the list. The constraint to be imposed is that *MONEY* must be different from all values in the accumulated list.

From the examples one can see that generalizing BAB to PS is straightforward. The notion of currently best solution is replaced by currently accumulated information:

- Initial accumulated information.
- A procedure that combines the previously accumulated information and a solution and returns the newly accumulated information.
- A procedure that takes the accumulated information and computes a constraint to be imposed. This replaces `Constrain` in BAB.

Otherwise the engine for PS is identical to the engine for BAB with explicit state as presented in Section 6.4. The formulation for BAB without explicit state cannot be used for PS, since it relies on the identity of solutions.

An interesting application of PS is symmetry elimination during search. PS has been used by Backofen and Will for symmetry elimination [8], which has been successfully applied to the prediction of protein structures [7].

7. Recomputation

This chapter introduces recomputation as an essential technique for search engines. Recomputation saves space, possibly at the expense of increased runtime. Recomputation can also save time, due to an optimistic attitude to search. Saving space and time makes recomputation an ideal candidate for solving large problems.

7.1 Overview

Search demands that nodes of the search tree must possibly be available at a later stage of exploration. A search engine must take precaution by either memorizing nodes or by means to reconstruct them. States are memorized by *cloning*. Techniques for reconstruction are *trailing* and *recomputation*. While recomputation computes everything from scratch, trailing records for each state-changing operation the information necessary to undo its effect. This chapter focuses on recomputation. Trailing and its relation to both cloning and recomputation are discussed in Section 14.2.

```
fun {Recompute S Is}
    case Is of nil then
        {Clone S}
    [] I|Ir then
        C={Recompute S Ir}
    in
        {Commit C I} C
    end
end
```

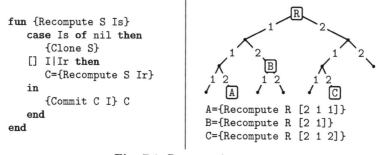

A={Recompute R [2 1 1]}
B={Recompute R [2 1]}
C={Recompute R [2 1 2]}

Fig. 7.1. Recomputing spaces.

The basic idea of recomputation with spaces is straightforward: any node in the search tree can be computed without search from the root node of the search tree and a description of the node's path. The procedure `Recompute` (Figure 7.1) recomputes a space from a space S higher up in the search tree

C. Schulte: Programming Constraint Services, LNAI 2302, pp. 59–67, 2002.

and a path between the two spaces represented as list of integers Is. The path is organized bottom-up, since it can be constructed easily that way during top-down exploration.

Indeterminism. Recomputation requires that a space can actually be recomputed. For a space S and a path Is the application {Recompute S Is} must always return equivalent spaces. This can go wrong due to indeterminism. The most likely source of indeterminism by multiple distributors is ruled out by design (Section 4.5.1). A second, albeit unlikely, source of indeterminism are the indeterministic constructs of Oz Light.

Indeterministic distributor creation is a programming error. The error typically proves fatal even without recomputation. Due to indeterminism, search is unpredictable and might take few milliseconds or several days.

Note that recomputation does not preclude randomly generated alternatives. A random generator is a *deterministic* program that on each invocation returns a number out of a pseudo-random sequence of numbers, for example [64, Chapter 3].

7.2 Full Recomputation

The most extreme version of recomputation is to always recompute spaces from scratch. The procedure DFE as shown in Figure 5.1 can be extended by two additional arguments: R for the root space and Is for the path of the current space S to the root.

Recursive applications of DFE additionally maintain the path to the root of the search tree. For example, the part of the search engine that explores the second alternative replaces cloning by recomputation and is as follows:

··· **then** C={Recompute R Is} **in** {Commit C 2} {DFE C R 2|Is}

To base exploration on recomputation alone is unfeasible. Suppose a complete binary search tree of height k (where a single node is assumed to have depth 0), which has 2^k leaves. To recompute a single leaf, k exploration steps are needed. Here and in the following the number of exploration steps is used as cost measure. An exploration step amounts to a commit-operation and the resulting propagation. This gives a total of $k2^k$ exploration steps compared to $2^{k+1} - 2$ exploration steps without recomputation (that is, the number of edges). Hence, full recomputation takes approximately $k/2$-times the number of exploration steps required without recomputation.

Last Alternative Optimization (LAO). Even though full recomputation is unfeasible, it allows to study a straightforward yet important optimization for depth-first exploration. After all but one alternative A of the root node N have been explored, further recomputation from N always starts with recomputing A. The optimization now is to do the recomputation step $N \rightarrow A$ only once. This optimization is well known. For example, it corresponds to the trust_me instruction in Warren's Abstract Machine [5, 154].

Let us consider a complete binary search tree of height k. The rightmost path in the tree has $k + 1$ nodes and requires k exploration steps (edges). A left subtree issuing from a node at height i on this path requires $i2^{i-1}$ exploration steps (this is the unoptimized case). Altogether, a tree of height k requires

$$k + \sum_{i=0}^{k} i2^{i-1} = 1 + k + (k - 1)2^k$$

exploration steps. Hence LAO saves approximately 2^k exploration steps.

7.3 Fixed Recomputation

The basic idea of combining recomputation with copying is as follows: copy a node from time to time during exploration. Recomputation then can start from the last copy N on the path to the root. Note that this requires to start from a copy of N rather than from N itself, since N might be needed for further recomputation.

A simple strategy is *fixed recomputation*: limit the number of steps needed to recompute a node by a fixed number m, referred to as *MRD (maximal recomputation distance)*. That is, after m exploration steps, a clone of the current node is memorized (sketched to the right for $m = 2$). Filled circles correspond to clones. The case of $m = 1$ coincides with no recomputation.

Analysis. Obviously, fixed recomputation decreases the memory needed during depth-first exploration by a factor of m. Suppose that the MRD is m and the height of the tree is k. The case for $k \le m$ corresponds to full recomputation. Suppose $k = lm$, where $l > 1$. Then each subtree of height m can be collapsed into a single 2^m-ary node. Each of the collapsed nodes requires $m2^m$ exploration steps. A 2^m-ary tree of depth $l - 1$ has

$$\sum_{i=0}^{l-1} (2^m)^i = \frac{2^{ml} - 1}{2^m - 1} = \frac{2^k - 1}{2^m - 1}$$

nodes. Altogether, a tree of depth k (for k being a multiple of m) needs the following number of exploration steps:

$$\frac{m2^m}{2^m - 1} \left(2^k - 1\right).$$

Hence fixed recomputation for a MRD of m takes

$$\frac{m2^{m-1}}{2^m - 1}$$

the number of exploration steps required without recomputation. The relative overhead is: for $m = 2$, 1.25, for $m = 5$, $80/31 \approx 2.6$, and for large m approximately $\frac{m}{2}$.

LAO. How LAO performs for an MRD of 2 is sketched
to the right. Nodes, where a clone is created during exploration, are black. Nodes, where a clone becomes available due to LAO, are gray. Unfortunately, the formulas

resulting from mathematical analysis have no straightforward solved form
and thus do not provide additional insight.

Exploration. The procedure DFRE (Figure 7.2(a)) implements depth-first exploration with fixed recomputation. S is the currently explored space, and R
is the space and Is the path for recomputation. The maximal recomputation distance is M (a free variable), whereas D is the current recomputation
distance. The shaded line implements LAO.

```
proc {DFRE S R Is D}
   case {Ask S}
   of failed then skip
   [] succeeded then raise [S] end
   [] alternatives(2) then C in
      if D==M then
         C={Clone S}
         {Commit S 1} {DFRE S C [1] 1}
         {Commit C 2} {DFRE C C nil M}
      else
         {Commit S 1} {DFRE S R 1|Is D+1}
         C={Recompute R Is}
         {Commit C 2} {DFRE C R 2|Is D+1}
      end
   end
end
```

(a) Exploration.

```
fun {DFRS P M}
   S={NewSpace P}
   proc {DFRE ···} ··· end
in
   try {DFRE S S nil M} nil
   catch [S] then [{Merge S}]
   end
end
```

(b) Search engine.

Fig. 7.2. Fixed recomputation.

Exploration maintains the following invariants:

− $1 \leq D \leq M$. If $D = M$, the invariant is maintained by cloning.

- If D < M, {Length Is} = D. If D = M, Is is either empty (due to LAO), or {Length Is} = D = M.
- A clone of S can be recomputed by {Recompute R Is}.

The full search engine is shown in Figure 7.2(b). It can be adapted to multiple alternatives as usual by binarization (Section 5.3). A straightforward optimization to speed up recomputation is to combine several commit-operations needed for binarization.

Other Search Engines. Recomputation can be incorporated straightforwardly into the search engines presented in Chapters 5 and 6. Note that LDS does not require recomputation, since the number of clones to be stored during exploration is limited by the typically small number of discrepancies.

The only search engine that requires some effort is BAB. Here recomputation must also take inject-operations into account (rather than only commit-operations). The discussion is postponed to Section 9.4, which introduces an abstraction for recomputation that naturally supports BAB.

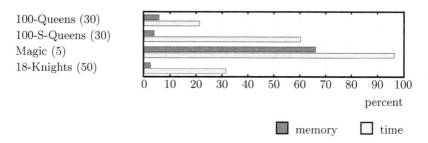

Fig. 7.3. Runtime and memory gain with fixed recomputation.

Empirical Results. Figure 7.3 shows empirical results of fixed recomputation for several example programs. All examples have in common that they are large: 100-Queens, 100-S-Queens, and 18-Knights have deep search trees; 100-Queens, Magic, and 18-Knights feature a large number of constraints and propagators. Detailed information on the examples can be found in Appendix A.1. As MRD for fixed recomputation the values given in parentheses are used.

The figures clearly show that fixed recomputation provides significant improvements with respect to runtime *and* memory requirements. It is worth noting that recomputation can save memory without runtime penalties even if the search tree is shallow (Magic).

Figure 7.4 relates the runtime to different MRDs for the 18-Knights problem. For a MRD from 1 to 10 the runtime is strictly decreasing because the time spent on copying and garbage collection decreases, while the plain run-

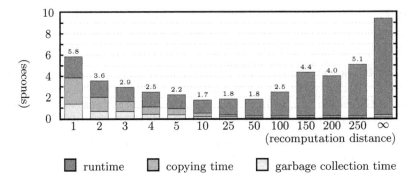

Fig. 7.4. Runtime for 18-Knights with fixed recomputation.

time remains constant. With further increase of MRD the runtime increases due to the increasing recomputation overhead.

Figure 7.4 shows a small peak at a MRD of 150. The search tree for 18-Knights has five failed nodes at a depth of around 260. This means that recomputation has to perform around 110 recomputation steps for each of the nodes. This phenomenon can be observed quite often: slight changes in the MRD (like from 100 to 150 for 18-Knights) results in unexpected runtime behavior. This indicates that for some parts of the search tree the assumption of recomputation is overly optimistic.

7.4 Why Recomputation Matters

Deep search trees are typical in solving large constraint problems. Large problems require a large number of decisions before arriving at a solution. A large number of decisions corresponds to a deep search tree.

The following simple facts are essential to understand why recomputation is an excellent technique for deep search trees and hence an excellent technique for solving large problems. Section 14.3 shows by an empirical comparison that recomputation outperforms all other constraint programming systems considered.

Space. Space is an obvious issue with deep search trees. Since space requirements are proportional to the tree's depth, the space required per node in the tree must be kept as small as possible.

Recomputation has the unique property that the space requirements are independent of the nodes and hence independent of the size of the problem. Space just depends on the tree's depth (only a list of integers).

Little Search. The size of a search tree grows exponentially with its depth. If a solution is found at all, only a small fraction of the search tree is explored.

Hence, the right attitude for exploring a deep search tree is to be *optimistic*: assume that a decision made is the right decision. *Cloning is pessimistic*: it assumes that each decision is likely to be wrong, since it always invests into cloning to undo the decision. *Recomputation is optimistic*: it assumes that every decision is right.

Clustered Failures. If exploration exhibits a failed node, it is quite likely that not only a single node is failed but that an entire subtree is failed. It is unlikely that only the last decision made in exploration has been wrong. This suggests that as soon as a failed node is encountered, the exploration attitude should become more pessimistic. This is addressed in the next section.

It is important to remember that efficient recomputation presupposes copying. Only their combination allows to select the ratio between optimism and pessimism.

7.5 Adaptive Recomputation

The analysis of fixed recomputation lead to the following two observations. Firstly, the optimistic assumption underlying recomputation can save time. Secondly, the fixed and hence possibly erroneous choice of the MRD can inhibit this.

The following strategy is simple and shows remarkable effect, since it honors the "clustered failures" aspect. During recomputation of a node N_2 from a node N_1, an additional copy is created at the middle of the path from N_1 to N_2. This strategy is referred to as *adaptive recomputation*.

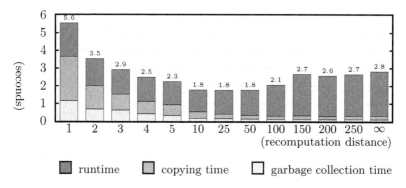

Fig. 7.5. Runtime for 18-Knights with adaptive recomputation.

Runtime. Figure 7.5 shows the runtime for adaptive recomputation applied to 18-Knights. Not only the peak for a MRD of 150 disappears, also the runtime for large MRD values remains basically constant. Even if copies are created during recomputation only (that is the MRD is ∞) the runtime remains almost unaffected.

Memory. While adaptive recomputation is a good strategy as it comes to runtime, it does not guarantee that memory consumption is decreased. In the worst case, adaptive recomputation does not improve over copying alone.

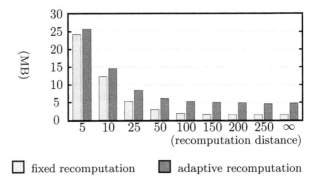

Fig. **7.6.** Memory requirements for 18-Knights.

Figure 7.6 shows the active heap memory for both fixed and adaptive recomputation applied to 18-Knights. The numbers exhibit that avoidance of peaks in runtime is not paid by peaks in memory (for MRDs between 1 and 5, memory requirements for both fixed and adaptive recomputation are almost identical and thus are left out).

For deep search trees the following technique saves memory. As soon as exploration has reached a certain depth in the search tree, it is quite unlikely that nodes high above are going to be explored. Hence, copies remaining in the upper parts of the tree can be dropped. This decreases memory consumption and does not affect runtime.

Adaptability. This is the real significance of adaptive recomputation: the choice of the recomputation distance is not overly important. Provided that the distance is not too small (that is, no excessive memory consumption), adaptive recomputation adjusts quickly enough to achieve good performance.

Figure 7.7 compares adaptive recomputation to no and fixed recomputation. The label $n\%$ means that the initial MRD is n percent of the total depth of the search tree. The comparison with no recomputation (Figure 7.7(a)) shows that adaptive recomputation offers almost always significant speedup. Additionally, it is clarified that the obtained speedup is almost independent of the initial choice of the MRD. This means that *adaptability* is really the most distinguished feature of adaptive recomputation.

On the other hand, adaptive recomputation performs almost as good as fixed recomputation with carefully hand-chosen MRDs (Figure 7.7(b)). This

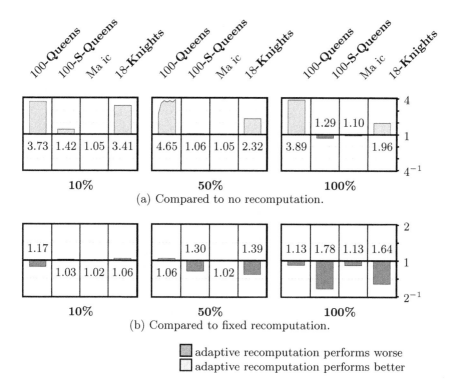

Fig. 7.7. Adaptability for different MRDs.

substantiates the claim that adaptive recomputation offers great potential even in case there is almost no knowledge about the problem to be solved. Starting with a rough guess on the initial MRD, adaptive recomputation behaves well. The runtime remains stable for a variation of the MRD by a factor of five (that is, between 10 and 50 percent of the total depth of the search tree).

8. Oz Explorer: Visual Search

The Oz Explorer is a graphical and interactive tool to visualize and analyze search trees. The Explorer is programmed from spaces. This chapter presents its motivation, design, and implementation.

8.1 Development of Constraint Programs

Development of constraint-based applications proceeds in two steps. The first step is to design a principally working solution. This is followed by the much harder task to make this solution scale to problems of real-world size. The latter task usually involves a high amount of experimentation to gain additional insight into the problem's structure. Meier reports in [79] that a large part of the development effort is spent on performance debugging. Therefore it is surprising that existing systems offer little support for the development of constraint programming applications.

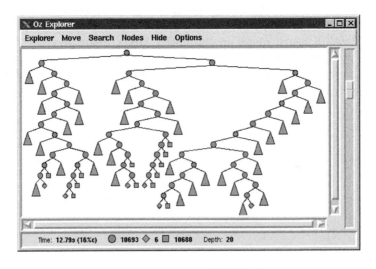

Fig. 8.1. Screenshot of the Explorer.

C. Schulte: Programming Constraint Services, LNAI 2302, pp. 69–78, 2002.
© Springer-Verlag Berlin Heidelberg 2002

This chapter presents the Oz Explorer as a visual constraint programming tool. It uses the search tree as its central metaphor (see Figure 8.1). The user can interactively explore the search tree which is visualized as it is explored. Nodes carry information on the corresponding constraints that can be accessed interactively. The Explorer can be used with any search problem, no annotations or modifications are required.

First insights into the structure of the problem can be gained from the visualization of the search tree. How are solutions distributed? How many solutions are there? How large are the parts of the tree explored before finding a solution? The insights can be deepened by displaying the constraints of nodes in the search tree. Is constraint propagation effective? Does the heuristic suggest the right alternatives? Interactive exploration allows following promising paths in the search tree without exploring irrelevant parts of it. This supports the design of heuristics and search engines.

Complex problems require a tool to be practical with respect to both efficiency and display economy. The amount of information displayed by the Explorer is variable: the search tree can be scaled and subtrees can be hidden. In particular, all subtrees without solutions can be hidden automatically.

The Explorer is one particular example of a user-guided interactive search engine that would not have been possible without first-class spaces.

8.2 Example: Aligning for a Photo

This section introduces the Oz Explorer by means of an example. Five people want to take a group photo. Each person can give preferences next to whom he or she wants to be placed on the photo. The problem to be solved is to find a placement that satisfies as many preferences as possible.

Figure 8.2 shows the script that models this problem. The record Pos maps the person's name to a position, that is, an integer between 1 and 5. All fields of Pos are enforced to be distinct by the propagator FD.distinct. The list of preferences is mapped to a list Ful of 0/1 variables. An element is 1 in case the preference can be fulfilled or 0 otherwise. The overall satisfaction Sat is given by the sum of all elements of Ful. The positions Pos are distributed (by {FD.distribute naive Pos}) following a naive strategy.

Reified propagators are used to map preferences to 0/1 variables. A reified propagator employs a 0/1 control variable b. If the propagator is entailed (disentailed), then b is constrained to 1 (0). If b is 1 (0), the constraint of the reified propagator is enforced (its negation is enforced). The reified propagator Pos.A+1=:Pos.B (Pos.B+1=:Pos.A) expresses that A is placed to the left (right) of B. Thus, the control variable of the reified propagator stating that the sum of both is 1, yields 1 if A and B are placed next to each other, and 0 otherwise.

```
Names = [a b c d e]
Prefs = [a#c b#e c#d c#e d#a d#e e#a e#b]
proc {Photo Sol}
    Pos = {FD.record pos Names 1#{Length Names}}
    Ful = {Map Prefs fun {$ A#B}
                        (Pos.A+1 =: Pos.B) + (Pos.B+1 =: Pos.A) =: 1
                    end}
    Sat = {FD.int 0#{Length Prefs}}
in
    {FD.distinct Pos}
    {FD.sum Ful ´=:´ Sat}
    Sol = sol(pos:Pos ful:Ful sat:Sat)
    {FD.distribute naive Pos}
end
```

Fig. 8.2. Program to solve the photo alignment problem.

The Explorer is used to search for a best solution to the `Photo` problem. The optimality criterion is described by a binary procedure stating that the satisfaction must increase with the solutions found:

```
{Explorer script(Photo proc {$ Old New}
                            Old.sat <: New.sat
                    end)}
```

The Explorer shows a single distributable node. Prompting for the next solution explores and draws the search tree up to the first solution as shown to the right. Exploring and drawing the search tree can be stopped at any time and resumed later at any node. This is important for problems which have large or even infinite subtrees in its search tree.

Double-clicking the solution displays the constraints of the succeeded space using the Oz Browser (a concurrent tool to visualize basic constraints) [108]. The first solution is as follows:

```
sol(pos: pos(a:1 b:2 c:3 d:4 e:5)
    ful: [0 0 1 0 0 1 0 0]
    sat: 2)
```

As understanding textual output can be difficult, the Explorer can employ user-defined display procedures. Suppose a procedure `DrawPhoto` that displays constraints graphically. The Explorer is configured such that double-clicking a node applies `DrawPhoto` to the node's constraints by

```
{Explorer add(information DrawPhoto)
```

Figure 8.3(a) shows a particular instance of graphical output for the previously found solution. An arrow between names shows a fulfilled preference, whereas the circled number above a name yields the number of non-fulfilled preferences of that person.

Invoking search for all solutions yields the search tree as shown in Figure 8.4(a). The best solution is the rightmost succeeded node.

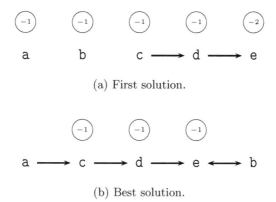

(a) First solution.

(b) Best solution.

Fig. 8.3. User-defined display for solutions of `Photo`.

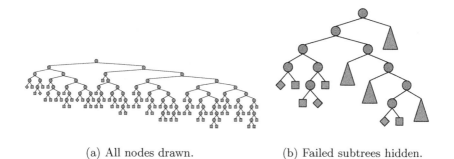

(a) All nodes drawn.   (b) Failed subtrees hidden.

Fig. 8.4. Search tree for `Photo`.

Although `Photo` is a simple problem, it is hard to find solutions and paths leading to them. The Explorer provides support to hide all subtrees which contain only failed leaves by drawing these subtrees as triangles. After applying this functionality, the search tree looks as shown in Figure 8.4(b).

By double-clicking the rightmost solution (the Explorer assists in finding certain nodes by moving a cursor to it), the best solution is displayed as shown in Figure 8.3(b).

The Explorer reports in its status bar that the entire search tree has 72 distributable, 3 solved, and 70 failed nodes. The tree indicates by the length of paths leading to failed leaves that the alternatives do not result in much constraint propagation. A better distribution heuristic should lead to more constraint propagation. The amount of constraint propagation depends on how many propagators are triggered to amplify the constraint store. So it is better to assign a value to a variable on which many propagators depend.

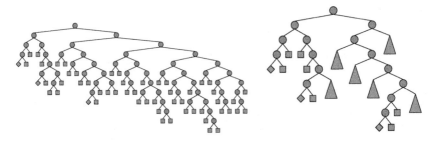

(a) All nodes drawn.   (b) Failed subtrees hidden.

Fig. 8.5. Search tree for `Photo` (with improved distribution strategy).

This is done by replacing the shaded distribution strategy in Figure 8.2 by a strategy implementing the idea from above:

`{FD.distribute generic(order:nbSusps) Pos}`

The Explorer is applied to the modified problem to study the impact on the search tree. The resulting tree is shown in Figure 8.5. The Explorer's status bar displays that the tree now has 54 distributable nodes, 3 solution nodes, and 52 failed nodes. That is, the number of nodes has decreased by about 25%. From the search tree one can conclude that it is much harder to prove optimality of the last solution than to actually find it.

The search tree in Figure 8.5 reveals that the third and fourth large subtree have the same shape. A common reason for subtrees exactly looking alike is that search aims at symmetrical solutions. By using the Explorer to access constraints of nodes in the right part of the tree, it becomes apparent that search is aiming at solutions symmetrical (that is, with people placed in reverse order) to those in the tree's left part. The search tree can be reduced in size by removing these symmetries. Some of them can be removed by

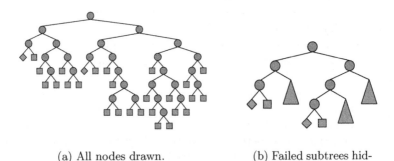

(a) All nodes drawn.   (b) Failed subtrees hidden.

Fig. 8.6. Search tree for `Photo` (with some symmetries removed).

placing two persons, say the first and the second in the list of persons, in a fixed order. Hence, the following constraint is added to the program:

    Pos.{Nth Names 1} >: Pos.{Nth Names 2}

Applying the Explorer to the new problem and searching for all solutions draws the search tree as in Figure 8.6(a). The tree now has only 27 distributable nodes, 2 solution nodes, and 26 failure nodes. Thus, removing just these symmetries reduces the number of nodes by 50%. Figure 8.6(b) displays the tree after hiding all failed subtrees.

8.3 Features

The main features of the Explorer are as follows.

Direct Use and Manipulation. Using the Explorer does not require any modification of the script. After having applied the Explorer to the script, all actions can be invoked by mouse-clicking, menu-selection, or keyboard accelerators.

Interactive and Incremental Exploration. Search can be used in an interactive fashion: the user can explore any part of the search tree step-by-step. Promising paths in the search tree can be followed without being forced to follow a predefined strategy. Furthermore, depth-first exploration of the search tree for one solution or for all solutions is supported. The Explorer is fully incremental: exploration of the search tree can be stopped at any time and can be resumed at any node.

Ergonomic Visualization. After creation of the search tree, the Explorer computes a layout for the newly created part of the search tree and updates the drawing of the tree. The drawn tree can be scaled by direct manipulation of a scale bar. Any subtree of the search tree can be hidden by replacing it with a small triangle. Special support is provided to hide subtrees which contain failed leaves only. By visualizing the search tree, one can gain insights into the search process. How are the solutions distributed? Is a first solution found without too many failed nodes? Is it hard to prove optimality of the last solution found? The possibility of hiding failed parts of the search tree assists finding relevant paths leading to solutions.

User-Defined Access to Constraints. All but the failed nodes carry as information their spaces. Each node's space can be displayed with user-defined or predefined display procedures. It is possible to compare the spaces attached to any two nodes, which assists to understand how the nodes differ.

Statistics Support. The Explorer provides brief statistical information in a status bar. Additionally, it is possible to display statistical information for each subtree. User-defined procedures can be used to process and display the statistical information. For instance, a bar chart showing how many failures occur between solutions can help to understand how hard it is to prove optimality in best-solution search.

A user manual that includes the description of an API (application programming interface) for the Explorer is [124].

8.4 Implementation

The Explorer manipulates a search tree that is implemented as a tree of objects. Each node is an object which stores the corresponding space. The object's class depends on the space to be stored, that is, whether the space is failed, succeeded, or distributable.

The implementation is factored into the following three parts:

User Interface. The user interface controls the invocation of operations on the search tree. Invoking an operation at the user interface sends a message to the object and leads to execution of the corresponding method.

Layout and Drawing. The methods for computing the layout use an incremental version of the algorithm presented in [63]. The graphical part of the user interface and the drawing of the tree uses the object-oriented graphics interface to Tcl/Tk [99] available in Oz [125]. I first considered using existing tools for computing and drawing layouts for graphs (for example, VCG [115] and daVinci [36]). Unfortunately, it is hard to design a powerful user interface, since the tools come with a user interface on their own that allows for limited customization only. More severely, they fail to support efficient incremental updates.

Exploration. Construction of the search tree is started with creating the root node. Further nodes are created as exploration proceeds.

The Explorer uses recomputation for two different purposes. Firstly, recomputation is used during exploration as in Chapter 7. In contrast to other search engines discussed so far, the Explorer keeps the entire explored part of the search tree. The search tree is kept for visualization but also to allow access to the corresponding spaces. For this purpose, recomputation is absolutely necessary, since keeping an exponential number of spaces is infeasible. The recomputation scheme employed is similar to that of fixed recomputation (Section 7.3) so that only nodes at a certain depth store a space, all other are recomputed on demand.

A useful optimization is to always recompute spaces of nodes occurring in subtrees that do not contain a solution[1]. This is motivated by the fact that the focus of interest is usually on nodes that are solutions or that lead to solutions.

---

[1] This technique has been suggested by Joachim P. Walser.

8.5 Evaluation

This section compares runtime and memory requirements of the Explorer
with that of non-visual search engines. Its purpose is to show that the Explorer is practical and scales to very large search trees. It demonstrates the
costs and benefits of some of the Explorer's features. The platform and examples used are described in Appendix A.

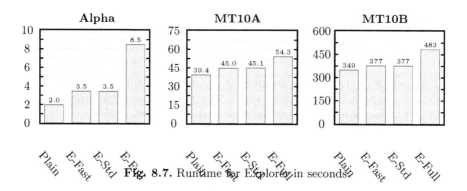

Fig. 8.7. Runtime for Explorer in seconds

Runtime. Figure 8.7 shows the runtime for the three example programs in
seconds. "Plain" is the runtime for a standard search engine without any
visualization features. The remaining numbers are taken with the Explorer.
"E-Fast" is the Explorer that uses full recomputation for state access and
hides failed subtrees while drawing. "E-Std" again hides failed subtrees and
uses a maximal recomputation depth of 5 for the state access. "E-Std" corresponds to the standard configuration of the Explorer. "E-Full" again uses
a maximal recomputation depth of 5 and draws the entire search tree.

For Alpha, using the Explorer introduces a runtime overhead of around
70%. This overhead is fairly modest, given that each exploration step is very
cheap. Drawing the entire tree is still feasible, although an overhead of approximately 300% is incurred.

The Explorer incurs for MT10A and MT10B approximately the same
overhead (MT10A: around 14%, MT10B: around 8%). Full drawing is still
feasible. The smaller overhead of MT10A and MT10B compared to Alpha is
due to the higher cost of each exploration step.

For all examples, "E-Fast" and "E-Std" show runtimes that can be regarded as equal. This means that creating additional copies during exploration to speed up state access is feasible with respect to runtime.

Memory. Figure 8.8 relates the memory requirements of the Explorer to
memory requirements of a non visual search engine. The meaning of "Plain"
through "E-Full" is as described before. The memory requirements of the
underlying graphics toolkit are excluded, they are mentioned below.

The important points for the Explorer's memory requirements are:

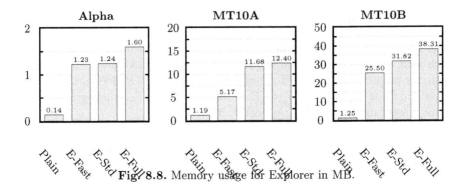

Fig. 8.8. Memory usage for Explorer in MB.

- The memory requirements are modest, even for a standard PC.
- Full recomputation for state access can have remarkable impact. For MT10A, the required memory is decreased by more than 50%.
- Memory requirements for the Explorer are independent of the memory requirements of the particular problem. When using full recomputation, the memory required by the Explorer depends only on the size and shape of the search tree. Only when creating additional copies to speed up state access, the memory requirements depend on problem size.
- Drawing the full tree has no strong impact on the memory requirements of the Explorer itself.

  This is different when also considering the memory requirements of the underlying graphics engine (Table 8.1). For all but MT10B with "E-Full" the memory requirements remain modest. Drawing the full search tree for MT10B is a borderline example. While a standard personal computer with 256 MB can handle this (no swapping occurs, this is witnessed through the modest runtime), full exploration for bigger examples is out of reach.
- Hiding failed subtrees is not only essential for arriving at an understanding of the search tree. It is also an excellent technique to keep the memory requirements low.

The runtime and the memory requirements can be summarized as follows. The Explorer is perfectly capable of exploring and visualizing very large search trees. Recomputation makes the memory requirements problem independent and makes the Explorer capable of handling large problems with

Table 8.1. Approximate memory usage of graphics engine in MB.

Example	E-Std	E-Full
Alpha	≈ 1.5	≈ 5
MT10A	≈ 3	≈ 5.5
MT10B	≈ 4	≈ 105

large search trees. Features such as recomputation for state access and hiding of failed subtrees are essential for the scalability of the Explorer.

8.6 Related Work

In the following the Explorer is related to the Grace tool [79], which is built on top of the Eclipse Prolog system [3]. The Grace tool is intended to support the development and debugging of finite domain constraint programs. Rather than using the metaphor of a search tree, it maintains and displays a backtracking history of the finite domain variables involved.

Exploration of the search space is not user-guided but fixed to a depth-first strategy. In contrast to the Explorer, it allows tracing of constraint propagation. The display of information supports different levels of detail, but cannot be replaced by user-defined display procedures. To use the Grace tool the user's program requires modification.

Similar in spirit to Grace is the CHIP search tree tool [130] which has been inspired by the Explorer. The strength of this tool lies in the visualization of finite-domain constraint propagation and in particular the visualization of global constraints. As with Grace, the CHIP search tree tool does not support interactive exploration of the search tree.

The Oz Explorer is focused on search only and does not address the visualization of constraint propagation. Instead, the Explorer relies on other tools for that purpose. In the context of Oz, the Oz Investigator offers this functionality [83].

In the area of parallel logic programming, tools are used to visualize the parallel execution of programs, for example, the Must Tool [138, 62] and the VisAndOr Tool [16]. These tools visualize the (OR-parallel) search process, however they are designed to be used off-line. During execution of a program a trace file is created. After execution has finished, the tool is used to visualize and analyze the created trace. This is very different from the Explorer, where exploration is interactive and user-controlled and where the user has access to the constraints of the search tree.

An overview on current research in the area of analysis and visualization tools for constraint programming and constraint debugging is [28].

9. Distributed Search

This chapter presents search engines that explore subtrees of a search tree in parallel. Parallelism is achieved by distribution across networked computers. The main point of the chapter is a simple design of the parallel search engine. Simplicity comes as an immediate consequence of clearly separating search, concurrency, and distribution. The obtained distributed search engines are simple yet offer substantial speedup on standard networked computers.

9.1 Overview

Search in constraint programming is a time consuming task. Search can be speeded up by exploring several subtrees of a search tree in parallel ("*or-parallelism*") by cooperating search engines called *workers*. To the right, exploration with three work-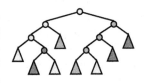
ers (the color of a subtree corresponds to the exploring worker) is sketched.

The chapter develops search engines that achieve parallelism by distributing workers across standard networked computers. The chapter has two main points. The first point is to provide a simple, high-level, and reusable design for parallel search. The second point is to obtain good speedup rather than good resource utilization.

Simple and Reusable Design. Parallel search is made simple by separating three issues: search, concurrency, and distribution.

Search. Workers are search engines that explicitly manipulate their state. The state corresponds to yet to be explored subtrees of the search tree. Explicit manipulation is mandatory since workers need to share subtrees. This has already been done in Section 5.5 for plain search and in Section 6.4 for best-solution search.

Concurrency. The main contribution of this chapter is the design of a concurrent search engine that adds communication and cooperation between workers. Communication and cooperation presupposes concurrency.

Distribution. How workers are distributed across networked computers is considered independently of the architecture of the concurrent engine. An important technique for sharing nodes across the network is recomputation.

C. Schulte: Programming Constraint Services, LNAI 2302, pp. 79–91, 2002.
© Springer-Verlag Berlin Heidelberg 2002

The approach obviously simplifies the design, since it allows to address concerns independently. It allows to reuse the concurrent architecture for other purposes, such as parallel execution on shared-memory multiprocessors and cooperative search for multi-agent systems.

The approach presupposes that search is encapsulated and combines well with concurrency and distribution. Since Oz is a concurrent language that supports distribution and since spaces are concurrency-enabled, the parallel search engines can be programmed entirely in Oz. The programming effort needed is around one thousand lines of Oz code.

Obtaining Speedup. Networked computers are cheap, ubiquitous, and mostly idle. Hence the criterion of success is whether a simple distributed search engine can offer substantial speedup. This differs from the traditional criterion of success for parallel search that aims at good utilization of specialized, expensive, and not widely available hardware.

A performance evaluation shows that the simple distributed engine offers substantial speedup already for small search trees. Large search trees as common for complex constraint problems provide almost linear speedup.

Related Work. There has been considerable work in the area of parallel search. Rao and Kumar discuss and analyze the implementation of parallel depth-first search in [113, 70]. Their focus is on the impact of the underlying hardware architecture and in particular how to best utilize the resources of the parallel architecture. Parallel execution on shared-memory multiprocessors and to a lesser extent on networked computers has received great attention in logic programming [19]. Early work that uses recomputation to distribute work is the Delphi Prolog system by Clocksin and Alshawi [21, 22].

Mudambi and Schimpf discuss in [93] distributed search that also relies on recomputation. A refinement of this work addresses branch-and-bound search [109]. Perron briefly sketches parallel search for ILOG Solver in [100]. All these approaches have in common that they are mostly focused on the description how each separate engine works. The discussion of the architecture by which the parallel engines communicate is missing or is at a low-level of abstraction. In contrast, this chapter is concerned with developing a high-level concurrent architecture underlying parallel search engines.

The approach to independently consider distribution and architecture is a consequence of the fact that distribution is provided orthogonally in Oz. Haridi et al. discuss this design approach in [46].

9.2 Distributed Oz

The basic idea of Distributed Oz is to abstract away the network as much as possible. This means that all network operations are invoked implicitly by the system as an incidental result of using particular language operations. Distributed Oz has the same language semantics as Oz Light by defining a distributed semantics for all language entities. The distributed semantics extends the language semantics to take into account the notion of *site* (or *process*). It defines the network operations invoked when a computation is distributed across multiple sites.

Partial Network Transparency. Network transparency means that computations behave the same independent of the site they compute on, and that the possible interconnections between two computations do not depend on whether they execute on the same or on different sites. Network transparency is guaranteed in Distributed Oz for most entities. While network transparency is desirable, since it makes distributed programming easy, some entities in Distributed Oz are not distributable.

There are two different reasons for an entity to be not distributable.

- The entity is *native* to the site. Examples are external entities such as files, windows, as well as native procedures acquired by dynamic linking. Native procedures depend on the platform, the operating system, and the process. Particular examples for native procedures in Mozart are most propagators which are implemented in C++ rather than in Oz [87].
- Distribution would be too complex. One class of entities for which distribution is too complex are computation spaces. Furthermore, even a distributed implementation of computation spaces would be of limited use, since a computation space typically contains native propagators.

Resource Access. For distributed computations that need to utilize resources of a distributed system, it is important to gain access to site-specific resources. Access is gained by *dynamic linking* of *functors* that return *modules*. Dynamic linking resolves a given set of resource-names (which are distributable) associated with a functor and returns the resources (which are site-specific).

A straightforward way to access site-specific resources is accessing them through active services. The service is distributable while its associated thread is stationary and remains at the creating site. Thus all resource accesses are performed locally. Services by this resemble remote procedure call (RPC) or remote method invocation (RMI).

Example 9.1 (Distributable Money). A definition of a functor for the *SEND+ MOST = MONEY* script as presented in Example 5.1 is as follows:

```
functor F
import FD export script:Money
define Figure 5.2
end
```

The functor F imports the FD module and returns a module that has a single field **script** that refers to the procedure **Money**. The functor F can be linked and its script can be executed as follows (DFS is introduced in Section 5.1):

```
{DFS {LinkFunctor F}.script}
```

Compute Servers. An Oz process can create new sites acting as *compute servers* [34]. Compute server creation takes the Internet address of a computer and starts a new Oz process with the help of operating system services for remote execution. The created Oz process can be given a functor for execution. Thus the functor gives access to the remotely spawned computations. Typically, a functor is used to set up the right active services and to get access to remote resources.

Further Reading. An overview on the design of Distributed Oz is [46]. A tutorial account on distributed programming with Mozart is [147]. The distributed semantics of logic variables is reported in [45]; the distributed semantics of objects is discussed in [148]. More information on functors, dynamic linking, and module managers in Mozart can be found in [35].

9.3 Architecture

The concurrent search engine consists of a single *manager* and several *workers*. The manager initializes the workers, collects solutions, detects termination, and assists in finding work for workers. Workers explore subtrees, share work with other workers, and send solutions to the manager.

9.3.1 Cooperation

Manager and workers are understood best as concurrent autonomous agents that communicate by exchanging messages. The architecture of the concurrent search engine is sketched in Figure 9.1.

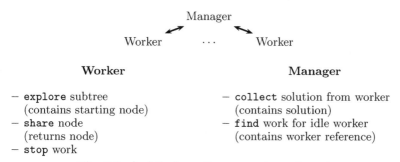

Fig. 9.1. Architecture of concurrent search engine.

Initialization. The concurrent search engine is initialized on behalf of the manager. The manager sends an **explore**-message for the root node of the search tree to a single worker. This single worker then starts working by exploring. A worker that currently explores a subtree is *busy* and *idle* otherwise.

Exploration. A worker works by exploring nodes of the search tree. By working it generates new work (new nodes).

Finding Work. Suppose that worker W_i is idle. It announces this fact to the manager by sending a find-message. The manager then tries to find a busy worker W_b that is willing to share work with W_i. If the manager finds work, it informs W_i by sending an explore-message containing the work found. To allow communication back from the manager to W_i, the find-message contains a reference to W_i.

The manager maintains a list of possibly busy workers which are not known to be idle, since the manager has not received a find-message from them. From this list the manager picks a worker W_b and then sends a share-message to W_b.

When W_b receives a share-message, it first checks whether it has enough work to fulfill the request. A worker receiving a share-message can be unable or unwilling to share work. It can be unable, because it is idle. It can be unwilling, because it has so little work left such that sharing it might make the worker idle itself (for example, the worker has only a single node left). In case the worker is willing to share work, it removes a node from its own pool of work and sends it to the manager. When the manager receives the node, it forwards the node to the requesting worker.

If the manager is informed that a share-message has been unsuccessful, it tries the next busy worker. If all busy workers have been tried, it starts over again by re-sending the initial find-message.

Collecting Solutions. When a worker finds a solution, it sends a collect-message containing the solution to the manager.

Termination Detection. The manager detects that exploration is complete, when the list of presumably busy workers becomes empty.

Stopping Search. If the search tree needs partial exploration (for example, single-solution search) the manager can stop search by sending a stop-message to all workers.

Almost all communication between manager and workers is asynchronous. The only point where synchronization is needed, is when the manager decides whether finding work has been successful. This point is discussed in more detail in Section 9.3.3.

Important Facts. The concurrent search engine does not loose or duplicate work, since nodes are directly exchanged between workers. Provided that the entire tree is explored, the number of exploration steps performed by the concurrent engine is the same as by the standard depth-first engine.

The exploration order is likely to be different from left-most depth-first. The order depends on the choice of the nodes to be exchanged between workers and is indeterministic. For all-solution search this has the consequence that the order in which the manager collects solutions is indeterministic. For single-solution search this has the consequence that it is indeterministic which solution is found. In addition, it is indeterministic how many exploration steps are needed. The number can be smaller or greater than the number of exploration steps required by depth-first exploration. The phenomenon to require less steps is also known as *super-linear speedup*.

9.3.2 Worker

A worker is a search engine that is able to share nodes and that can be stopped. Figure 9.2(a) summarizes which messages a worker receives and sends. The ability to share work requires explicit state representation (Section 5.5). A worker knows the manager and maintains a list of nodes that need exploration ("work pool").

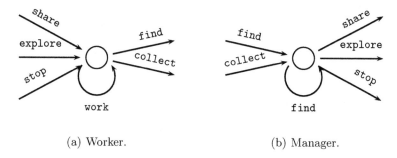

(a) Worker.   (b) Manager.

Fig. 9.2. Summary of messages.

Concurrent Control. The worker is implemented as active service (Section 3.3.3). It runs in its own thread and sequentially serves the messages it receives. This simple design is enough to ensure consistency of the worker's state in a concurrent setting.

The worker recursively invokes exploration as sketched in Section 5.5 by sending an exploration message to itself. By message sending, exploration and communication with the manager is easily synchronized.

Which Node to Share. A promising candidate is the highest node in the search tree, since it is likely that the subtree issuing from it is large (*"large work granularity"*). A large subtree prevents that the requesting worker becomes idle soon and thus helps to avoid excessive communication. Later it will become clear that sharing the highest node is a particularly good choice for distribution.

9.3.3 Manager

The manager is implemented as an active service, the messages it sends and receives are summarized in Figure 9.2(b). The manager knows all workers.

They are needed for initialization and for stopping. The manager maintains a list of workers not known to be idle and a list of solutions.

Finding Work. Finding work can be a time consuming task since it can take several attempts to seek for a worker that is able to share work. Hence, it is infeasible to block the manager while seeking work.

A design that does not block the manager is as follows. When the manager receives a `find`-message, it spawns a new thread that takes the current list of busy workers as snapshot. Directly after thread creation, the manager is again available to serve incoming messages. If no work has been found, the initial `find`-message is sent again to the manager and the thread terminates. This is repeated until either work is found or no presumably busy workers are left.

The solution to take a snapshot of the currently busy workers upon message receipt is simple but has the following drawback. The manager might ask workers that are still contained in the snapshot but have already announced that they are idle themselves. This can result in a delay of the manager to find work and thus the initially requesting worker might remain idle for a longer period of time.

9.3.4 Best-Solution Search

The main design issue in best-solution search is how to maintain the so-far best solution. The sequential branch-and-bound engine always knows the so-far best solution (BS in Figure 6.2). This is difficult to achieve in a concurrent setting with several workers. Maintaining the best solution for each worker would require a large communication and synchronization overhead. Instead, a design is preferred, where both manager and workers maintain the so-far best solution as follows:

Manager. When the manager receives a new solution through a `collect`-message, it checks whether the solution is really better. If the solution is better, the manager sends it to all workers. This requires a `better`-message that contains the so-far best solution.

Worker. When a worker finds a new solution, it stores the solution as so-far best solution and informs the manager by sending a `collect`-message. When a worker receives a `better`-message, it checks whether the received solution S_1 is better than its so-far best solution S_2.

Note that it is correct albeit inefficient, if the worker does not check whether the received solution S_1 is better. If S_1 is worse, S_1 will be replaced anyway, since the manager eventually sends a solution which is at least as good as S_2 (since it receives S_2 from this worker). It might be better in case the manager has received an even better solution from some other worker.

The architecture sketched above entails that a worker might not always know the so-far best solution. This can have the consequence that parts of the search tree are explored that would have been pruned away otherwise. Thus the loose coupling might be paid by some overhead. This overhead is referred to as *exploration overhead*.

The worker is based on the branch-and-bound search engine with explicit state as presented in Section 6.4.

9.4 Distributed Search Engines

This section discusses how to adopt the concurrent search engine such that its workers are distributed across networked computers.

Search Engine Setup. The setup of the search engine uses compute servers. The manager is created first. Then a new Oz process is created for each worker. Typically, each process is created on a different networked computer. In case a computer has more than a single processor, it can make sense to create more than a single process on that computer.

Each newly created process is given a functor that creates the worker service. It is important that the functor can be given first-class, since the worker requires access to the manager service. Applying the functor returns reference to the now created worker.

Distributing Nodes. Since spaces are not distributable, workers cannot exchange work by communicating spaces directly. Scripts are not distributable, since they typically contain references to native propagators. However, a functor that on application returns the script *is* distributable. This means that the root space can be recomputed via the script from the given script-functor by all workers.

Given the root space, work can then be communicated by communicating paths in the search tree that describe how to recompute nodes:

$$\text{node} \quad \longleftrightarrow \quad \text{root} + \text{path}$$

When a worker acquires new work, the acquired node is recomputed. This causes overhead, referred to as *recomputation overhead*. The higher the node in the search tree, the smaller the recomputation overhead. For this reason, sharing the topmost node is a good choice. Since all nodes are subject to sharing, a worker must always maintain the path to recompute a node.

Recomputable Spaces. In the following, *recomputable spaces* (r-spaces for short) are employed as convenient abstractions for distributed search engines. An r-space supports all space operations. Additionally, an r-space provides an *export* operation that returns the path for recomputation. Search engines that employ r-spaces rather than "normal" spaces are otherwise identical, since r-spaces provide the same programming interface.

The key feature of an r-space is that commit-operations are executed lazily on demand. Lazy execution is beneficial for two reasons. Firstly, not the entire search tree might be explored during single solution search (this point is discussed in Section 5.5). Secondly, a node might be handed out to some other worker and thus might be wasted for the current worker.

An r-space encapsulates the following three components:

Sliding Space. It is initialized to a clone of the root space.
Pending Path. A list of pending commit-operations.
Done Path. A list of already done commit-operations.

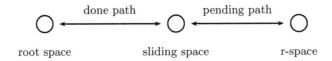

Fig. 9.3. Sketch of an recomputable space (r-space).

The sliding space always satisfies the invariant that it corresponds to a space that has been recomputed from the root space and the done path. This is sketched in Figure 9.3.

Initialization. Creation of an r-space takes a path P as input. The sliding space is initialized to a clone of the root space. The pending path is initialized to the path P. The done path is initialized to the empty path.
Commit. A commit to the i-th alternative adds i to the pending path.
Update. Updating an r-space performs all commit-operations on the pending path. Then the pending path is added to the done path and is reset.
Ask, Clone, Merge. Ask, clone, and merge update the r-space first and then perform the corresponding operation on the sliding space.
Export. Export returns the concatenation of done and pending path.

An r-space is extended straightforwardly to support best-solution search by storing a list of operations rather than a simple path. This list of operations contains elements of the form `commit(`i`)` and `constrain(`x`)`, where i is the number of an alternative and x is a solution. This presupposes that solutions are distributable.

The optimization that workers check whether received solutions are better (Section 9.3.4) helps to reduce the number of `constrain(`x`)`-elements on a path. Keeping the path short is important, since each operation on the path might be executed by multiple workers and even a single worker might execute each operation more than once.

Network Failure. What is not considered by now and left as future work is network failure. However, the interest is mostly on local area networks, where network failure is infrequent.

9.5 Evaluation

Evaluation uses common benchmark problems: Alpha, 10-S-Queens, Photo, and MT10 (Appendix A.1). They vary in the following aspects:

Search Space and Search Cost. All but MT10 have a rather small search space where every exploration step is cheap (that is, takes little runtime).

Strategy. For Alpha and 10-S-Queens all-solution search is used. For Photo and MT10 best-solution search is used.

Number of Solutions. 10-S-Queens has many solutions (approximately 10% of the nodes). This makes the example interesting, as each solution is forwarded to the manager. This assesses whether communication is a bottleneck and whether the manager is able to process messages quickly.

The choice of examples addresses the question of how good parallel search engines can be for borderline examples. MT10, in contrast, can be considered as a well-suited example as it comes to size and cost.

Total Overhead. Figure 9.4 shows the total overhead of a distributed search engine. The overhead is taken as the additional runtime needed by a distributed search engine with a single worker, where both worker and manager execute on the same computer compared to a sequential search engine. Information on software and hardware platforms can be found in Section A.3.

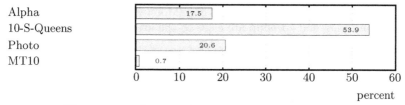

Fig. 9.4. Total overhead of distributed search engine.

The numbers suggest that for examples with small search space and small search cost, the overhead is less than 25%. This is due to the additional costs for maintaining r-spaces and message-sending. For large examples (MT10), the overhead can be neglected. The overhead of around 50% for 10-S-Queens is due to frequent communication between worker and manager. Compared to the small search cost, this overhead is quite tolerable.

Speedup. Figure 9.5 shows the speedup for a varying number of workers. All examples offer substantial speedup. For three workers all examples yield at least a speedup of two, and for six workers the speedup is at least around three. The speedup for MT10 with six workers is larger than 4.5.

For all combinations of workers and examples but 10-S-Queens with six workers the coefficient of deviation is less than 5% (in particular, for all

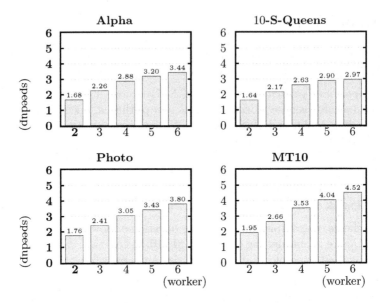

Fig. 9.5. Speedup.

combinations of MT10 less than 2%). For 10-S-Queens with six workers the coefficient of deviation is less than 10%. This allows to conclude that speedup is stable across different runs and that indeterminism introduced by communication shows little effect on the runtime. Moreover, this clarifies that both minimal and maximal speedup are close to the average speedup.

Work Granularity. Figure 9.6 shows the average work granularity which is amazingly coarse. Work granularity is the arithmetic mean of the sizes of subtrees explored by a worker in relation to the size of the entire tree. For all combinations of examples and workers the granularity remains close to ten percent. This means that the simple scheme for sharing work is sufficient.

Manager Load. A possible drawback of a single manager is the potential of a performance bottleneck. If the single manager is not able to keep up with processing find-messages, workers might be idle even though other workers have work to share. Figure 9.7 shows the load of the manager, where a load of 50% means that the manager is idle half of the runtime.

For all examples the manager has a load of less than 50%. For the more realistic examples Photo and MT10 the load is less than 15%. This provides evidence that the manager will be able to efficiently serve messages for more than six workers. There are two reasons why the load is quite low. Firstly, work granularity is coarse as argued above. Coarse granularity means that workers infrequently communicate with the manager to find work. Secondly, each incoming request to find work is handled by a new thread. Hence, the manager is immediately ready to serve further incoming messages.

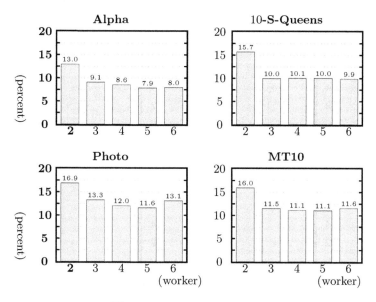

Fig. 9.6. Work granularity.

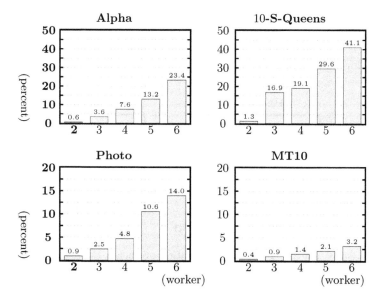

Fig. 9.7. Manager load.

Recomputation Overhead. Figure 9.8 shows the recomputation overhead, which is always less than 10%. This means that the price paid for distributing work across the network is low.

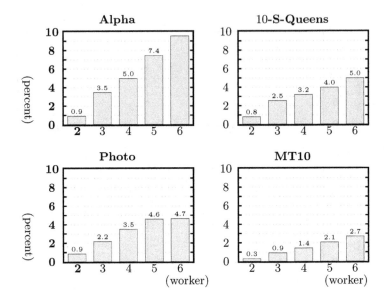

Fig. 9.8. Recomputation overhead.

Exploration Overhead. Exploration overhead occurs for branch-and-bound search and is due to the different order in which solutions are found (Section 9.3.4). Figure 9.9 shows the exploration overhead for Photo and MT10. The exploration overhead is almost exclusively the cause for the speedup loss.

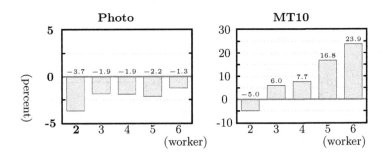

Fig. 9.9. Exploration overhead.

Exploration overhead is a consequence of performing branch-and-bound in parallel and is independent of the implementation of the search engines. A different approach to parallel best-solution search is presented by Prestwich and Mudambi in [109]. They use *cost-parallelism*, where several searches for a solution with different cost bounds are performed in parallel. This technique is shown to perform better than parallel branch-and-bound search.

10. Spaces for Combinators

This chapter extends computation spaces for programming composable constraint combinators. Composable means that combinators programmed from spaces can combine arbitrary computations, including computations already spawned by combinators.

10.1 Overview

Space-based programming of composable combinators requires that spaces are freely composable themselves. This is achieved by allowing spaces to be nested inside spaces, leading to a tree of spaces.

Example 10.1 (Negation with Spaces). This example considers the key issues that arise with programming combinators from spaces. Spaces localize failure through encapsulation. Hence, an obvious application for spaces seems to program a negation combinator for arbitrary statements. To be more concrete, the negation of X=Y is considered.

If encapsulated execution of X=Y fails, the negation of X=Y holds. If encapsulated execution of X=Y becomes stable, X=Y holds. However, due to the independence restriction introduced in Chapter 4, space creation waits until both X and Y become determined.

The independence condition prevents deciding failure early. If X and Y are aliased, speculative execution should already be able to detect failure. If X and Y are kinded to integers and have disjoint domains, speculative execution again should detect failure.

For early failure detection, spaces must be created immediately and constraints must be propagated into spaces immediately ("nested propagation").

Encapsulation must take variables situated in superordinated spaces into account. For example, speculative execution of X=Y aliases X and Y. The aliasing must be encapsulated inside the local space and must be invisible in the toplevel space (Figure 10.1).

Stability must take into account that a non-runnable space is not necessarily stable. For example, as soon as the thread executing X=Y terminates, the space is not runnable. However it is far from being stable: it can fail due to tells on X and Y.

C. Schulte: Programming Constraint Services, LNAI 2302, pp. 93–104, 2002.
© Springer-Verlag Berlin Heidelberg 2002

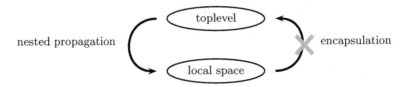

Fig. 10.1. Nested propagation and encapsulation for spaces.

The point to use spaces for combinators is to allow combination of arbitrary statements. Spaces are made composable so that statements that themselves employ combinators are eligible for further combination. A variety of combinators including a negation combinator are discussed in Chapter 11.

The chapter is concerned with the following aspects:

Space Tree. Spaces are organized in a tree that features nested propagation and encapsulation (Section 10.2).

Space Tree Manipulation. Space creation, cloning, and merging of spaces are extended to deal with the space tree. This in particular includes control conditions for the applicability of space operations (Section 10.3).

Control and Status. Stability is extended to capture nested propagation. A status mechanism that casts synchronization on spaces into synchronization on variables even supports debugging (Section 10.4).

The search-specific aspects of spaces such as distributor creation and committing to alternatives remain unchanged. The relation of spaces to the underlying programming language is discussed in Section 10.5.

10.2 Space Tree

As argued before, composable spaces lead to a *space tree*. The root of the space tree is the *toplevel space*. The direct predecessor S_1 of a space S_2 in the space tree is its *parent space*, and is written $S_1 \lessdot S_2$. Symmetrically, S_2 is a *child space* of S_1. The transitive closure of $\lessdot$ is denoted by $<$, and the transitive and reflexive closure by $\leq$.

Important subsets of the space tree with respect to a single space S are:

$$\uparrow S \quad := \quad \{S' \mid S' < S\} \qquad \Uparrow S \quad := \quad \uparrow S \cup \{S\} \quad = \quad \{S' \mid S' \leq S\}$$
$$\downarrow S \quad := \quad \{S' \mid S < S'\} \qquad \Downarrow S \quad := \quad \downarrow S \cup \{S\} \quad = \quad \{S' \mid S \leq S'\}$$

A space S_1 is *superordinated* to a space S_2, if $S_1 \in \uparrow S_2$. A space S_1 is *subordinated* to a space S_2, if $S_1 \in \Downarrow S_2$. Note that a space is subordinated but not superordinated to itself.

The space tree is not to be confused with the search tree. Spaces that implement nodes of a search tree are typically created by cloning. As will become clear in the following section, spaces of a search tree are siblings in the space tree.

Space Constituents. The constituents of a space and notions such as situated entity and home space remain unchanged. In the following, S_C refers to the current space.

Freshness and Visibility. As before, the set of variables and names for each space are disjoint. Variables and names are visible in all spaces that are subordinated to their home. That is, computations in space S can potentially refer to variables and names in $\Uparrow S$.

Procedure Application. When a procedure application reduces in S, the appropriate procedure is taken from the union of procedure stores in $\Uparrow S$. As a consequence of the disjointness of names, the procedure to be applied is uniquely determined.

Tell. In order to capture nested propagation, execution of a tell statement

$$x = v$$

tells $x = v$ in *all* spaces $\Downarrow S_C$. This ensures an important monotonicity invariant: if $S_1 \lessdot S_2$ and ϕ_i is the constraint of S_i, then ϕ_2 entails ϕ_1 ("children know the parent's constraints"). The invariant holds since children initially inherit the parent's constraints (Section 10.3.1).

Failure. An unsuccessful attempt to tell $x = v$ *fails* S_C. Failing S_C stops all computations in $\Downarrow S_C$ as follows: all threads in $\Downarrow S_C$ and all spaces $\downarrow S_C$ are discarded.

Sendability. Sendability as defined in Section 4.6 disallows undetermined variables in messages. Sendability is liberalized as follows. A variable x is *sendable* from S_1 with store ϕ to S_2, if $S_2 \in \Uparrow S_1$ and: there is no variable y with $x \vartriangleleft_\phi y$ and $S_2 \lessdot \mathcal{H}(y)$, and there is no name ξ with $x \vartriangleleft_\phi \xi$ and $S_2 \lessdot \mathcal{H}(\xi)$ ($\vartriangleleft$ is introduced in Section 3.2.1).

Example 10.2 (Space Tree). Consider the space tree in the top-left of Figure 10.2. The toplevel space is S_0, the spaces S_1 and S_2 are children of S_0, and S_{21} is child of S_2. The variable x is situated in S_0 and the variable y is situated in S_2.

The variable x is visible in all spaces, and y is visible in S_2 and S_{21}. The variable x is sendable to S_0 from all spaces, and y is sendable from S_{21} to S_2 but not to S_0.

- Telling $x = b$ in S_2 also tells $x = b$ in $S_{21} \in \Downarrow S_2$.
- Telling $x = b$ in S_0 tells $x = b$ in all spaces. Space S_1 is failed by the tell.
- Telling $y = a$ in S_{21} affects S_{21} only. Now, y is sendable from S_{21} to S_0, but still not sendable from S_2 to S_0.

10.3 Space Tree Manipulation

This section is concerned with space creation, cloning, and merging.

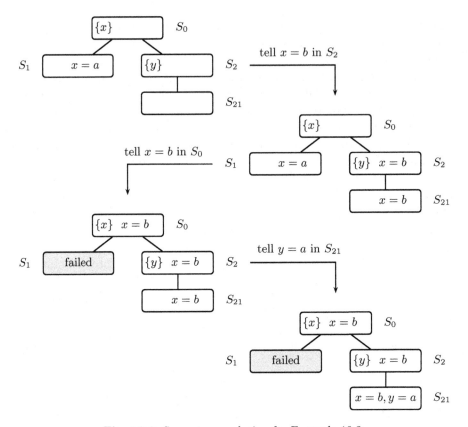

Fig. 10.2. Space tree evolution for Example 10.2.

10.3.1 Space Creation

A new space is created by

$$\{\texttt{NewSpace}\ x\ y\}$$

Reduction blocks until x becomes determined. A new name ξ is created with home S_C. A new space S is created as child of S_C the following way:

- The *root variable* of S is initialized with a fresh variable z with home S.
- The set of local variables is initialized to contain z. The set of local names and the procedure store are initialized as being empty.
- The constraint store is initialized with the constraints of S_C. This ensures the invariant that a child's constraint always entails the parent's constraint.
- A thread is created in S to execute $\{x\ z\}$.

Finally, the statement $y = \xi$ is pushed.

Visibility of Spaces. Due to the script's free variables, computations in S can potentially access all variables situated in $\Uparrow S$. And by construction, all children of S can be referred to in S. Altogether, computations in S (black to the right) can possibly refer to any space (gray) in

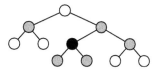

$$\mathcal{V}(S) = \{S_2 \mid S_1 \in \Uparrow S \, , \, S_1 \stackrel{\cdot}{<} S_2\}$$

Note that $\mathcal{V}(S)$ includes S and excludes the toplevel.

Space Access is Explicit. An important invariant in the design of first-class spaces is that reference to a space is *explicit*. The only way to gain first-class access to a space is by passing references obtained by `NewSpace`. This also entails that there is no first-class reference to the toplevel space.

A different design would be to allow implicit access by a primitive `{ThisSpace x}` that returns a reference to the current space. Implicit access would render abstractions programmed from spaces unsafe. Computations controlled by space-based abstractions could gain access to the current space and could break the abstraction's invariants. On top of that, implicit access would allow to gain access to the toplevel. Having no first-class access to the toplevel space simplifies the design considerably. Otherwise, most operations need to take care of the toplevel space as a special case.

Cloning Spaces. Cloning also creates new spaces and needs to take into account the space tree. Firstly, cloning a space S includes cloning all spaces in $\Downarrow S$. Secondly, the parent of the clone is the current space, which makes cloning similar to space creation.

10.3.2 Merging Spaces

Reduction of

$$\{Merge \ x \ y\}$$

synchronizes on x being $\xi \mapsto S$. Reduction considers the following cases:

- If S is failed, the current space S_C is failed.
- If S is merged, an exception is raised.
- If S is not admissible (explained below), an exception is raised.
- If both S_C and S contain a distributor thread, an exception is raised. This maintains the "at most one distributor" invariant (Section 4.5.1).

Otherwise, S is merged with S_C as follows:

- S is marked as merged.
- The set of local variables (names) of S_C is updated to include the local variables (names) of S. The invariants discussed in Section 10.2 exclude conflicts.

– Similarly, the procedure store of S_C is updated to include the mappings of S's procedure store. Again, no conflicts are possible.
– $y = z$ is pushed, where z is the root variable of S.
– All constraints of S are told in S_C.

Admissibility. Merging must obey a straightforward tree condition. Suppose that the space S to be merged is included in $\uparrow S_C$. By merging S, the space tree would evolve into a cyclic graph. Therefore, execution of $\{\texttt{Merge } x \; y\}$ such that x is $\xi \mapsto S$, raises an exception if $S_C \in \Uparrow S$. Note that even the case $S = S_C$ is excluded, since it is most likely a programming error worth being detected.

Spaces to which merging can be applied are *admissible*. To the right, the current space is black while the admissible spaces are gray. The set of admissible spaces with respect to the space S (typically, S is the current space) is defined as

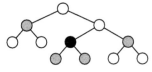

$$\begin{aligned} \mathcal{A}(S) \quad &:= \quad \mathcal{V}(S) - \Uparrow S \\ &= \quad \{S_2 \mid S_1 \in \Uparrow S \, , \; S_1 \dot{<} S_2\} - \Uparrow S \end{aligned}$$

Admissibility is a very general condition. This has the advantage that it can be used as single control condition for all operations on spaces. The following example shows that admissibility for merging is indeed useful.

Example 10.3 (Downward Merge: Partial Evaluation). How to use cloning and merging of spaces for partial evaluation is shown in Example 4.8.

The essence of using spaces for partial evaluation is to compute a space and to use it multiply by merging a clone of it. Typically, the clone S_1 is a child of the toplevel space S. The clone S_1 is merged to a space S_2 which is subordinated to S but not to S_1. Figure 10.3 sketches the space tree.

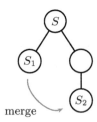

merge

Fig. 10.3. Downward merge for partial evaluation.

Merged Spaces are Not Transparent. The attempt to perform an operation on a merged space raises an exception. A different design would make merged spaces transparent: after merging S with S_C, any reference to S is automatically redirected to S_C instead (similar to logic variables). This design, however, would make space access implicit. In particular, `ThisSpace` could be programmed:

```
fun {ThisSpace}
   S={NewSpace proc {$ _} skip end} in _={Merge S} S
end
```

10.3.3 Injecting into Spaces

The operation

$$\{\texttt{Inject } x \; y\}$$

with x being $\xi \mapsto S$ is also restricted in that S must be admissible. Additionally, if y refers to a procedure with home S' and $S' \notin {\Uparrow}S$, an exception is raised.

Merged Spaces are Still Not Transparent. The hypothetical design that makes merged spaces transparent would allow to express `Inject` from `NewSpace` and `Merge`:

```
proc {Inject S2 P}
   S2={NewSpace proc {$ X} {Merge S1 X} {P X} end}
end
```

Here `S1` and `S2` would refer to the same space after merging `S1` with `S2`. Since merged spaces are not transparent, `Inject` is primitive. It is possible to create a space with the right computations. However, the space has the wrong identity.

10.4 Control and Status

The motivating Example 10.1 outlined that a space that is not runnable is not necessarily stable: it still can be speculative in that the space might fail. There are two reasons why a space S can still fail, even though S is not runnable:

- The space S contains a thread T that synchronizes on x that is situated in ${\Uparrow}S$. In case a constraint is told on x in ${\Uparrow}S$, T is woken and makes S runnable.

  In this situation, T is *globally suspended* or *speculative*. If T suspends on a variable x with $\mathcal{H}(x) < S$, T *globally suspends for* $\mathcal{H}(x)$. The *global suspension set* $\mathcal{G}(T)$ is the set of variables on which T globally suspends.
- A variable x situated in $S' \in {\Uparrow}S$ is constrained in S. In case a constraint is told on x in S', S can fail. This can only be the case, if the constraint store of S is not entailed by the constraint store of S'. The constraint in S is *speculative*. Otherwise, the tell would already fail S' (and S would be discarded instead).

10.4.1 Stability

The intuition is that S is stable, if no tell in $\uparrow S$ can affect S. This is formalized as follows.

Runnable and Blocked. A space S is *runnable*, if $\Downarrow S$ contains a runnable thread. Otherwise, S is *blocked*. According to this definition, a failed space is blocked.

Stable and Suspended. A space S is *stable*, if S is blocked and remains blocked regardless of any tell statements executed in $\uparrow S$. A space is *suspended*, if it is blocked but not stable.

Succeeded and Distributable. A space is *distributable*, if it is stable and contains a distributor. A space is *succeeded*, if it is stable but neither failed nor distributable.

Entailed and Stuck. A space is *stuck*, if it is succeeded and contains a thread. Otherwise, a succeeded space is *entailed*. The distinction between entailed and stuck spaces is of great importance in Chapter 11.

Figure 10.4(a) summarizes the states of a space and their relationship. Recall that a space can also be semi-stable (Section 4.5.3). Semi-stability is an orthogonal issue and hence requires no discussion here. Figure 10.4(b) summarizes state transitions for spaces. A solid line represents a transition that occurs upon application of a space operation, the other "implicit" transitions are represented by dashed lines.

Stability of a space S is defined with respect to threads in $\Downarrow S$. One reason is that a thread in S can control and synchronize on computations in $\downarrow S$. A further reason is due to cloning: cloning synchronizes on stability and also clones subordinated spaces.

Since stability is defined with respect to trees of spaces, the following holds:

– If S is blocked, all spaces in $\downarrow S$ are blocked. Dually, if S is runnable, all spaces in $\uparrow S$ are runnable.
– If S is stable, spaces in $\downarrow S$ need not be stable. Dually, if S is suspended, spaces in $\uparrow S$ need not be suspended.

Stability captures synchronization that arises naturally with concurrent computations. Consider a speculative computation in S that processes data (that is, constraints) provided by some other concurrent computation. The space S becomes suspended if not all required data is provided. Only after all data is provided, S can become stable.

Stability has been first conceived by Janson and Haridi in the context of AKL [59, 44, 58]. Stability naturally generalizes the notion of entailment. Entailment is known as a powerful control condition in concurrent execution, which has been first identified by Maher [74] and subsequently used by Saraswat for the cc (concurrent constraint programming) framework [118, 117].

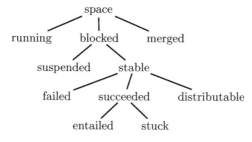

(a) Relation between states.

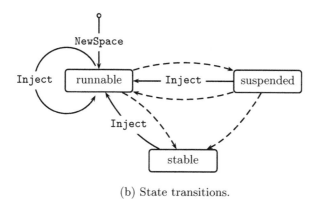

(b) State transitions.

Fig. 10.4. Summary of space states and transitions.

It is instructional to study stability and how stability interacts with failure and merging spaces by means of some examples.

Example 10.4 (Stability is Pessimistic). An important aspect of stability is that it is a pessimistic but safe and decidable approximation that a space is not speculative. Consider the following example

```
proc {Loop} {Loop} end
S={NewSpace proc {$ X} {Loop} end}
```

S is definitely not speculative, but never becomes stable. In the following example

```
local Y in S={NewSpace proc {$ X} Y=1 end} end
```

S never becomes stable, even though lexical scoping ensures that no tell on Y can fail S.

Example 10.5 (Local versus Global Variables). Suppose S is created by

```
local Y in S={NewSpace proc {$ X} X=Y end end
```

After the thread that executes X=Y terminates, S becomes stable: regardless of what is told for Y, S cannot become failed. This is in contrast to Example 10.1, where the constraint is speculative.

Example 10.6 (Failure and Stability). Consider the following example:

```
S1={NewSpace proc {$ S2}
                S2={NewSpace proc {$ X} Y=1 end}
            end}
```

where Y is a variable introduced in a space in ↑S1. Both S1 and S2 eventually become suspended, since S2 can be failed by a tell on Y. By telling Y=1, both spaces eventually become stable. By telling Y=2, S2 eventually becomes failed and S1 stable.

Example 10.7 (Merging and Stability). After execution of

```
S1={NewSpace proc {$ S2}
                Y in S2={NewSpace proc {$ X} Y=1 end}
            end}
```

S1 and S2 eventually become suspended. By

```
{Inject S1 proc {$ S2} {Merge S2 _} end}
```

S1 eventually becomes stable.

10.4.2 Status Variable

Ask as introduced in Section 4.4 synchronizes on stability of a space and then returns its status. A simpler design that casts synchronization on spaces to synchronization on variables is based on the idea of a *status variable*. As soon as a space reaches a stable state, information according to its status is told on the status variable. Ask is then programmed from a primitive that accesses the status variable.

Each space S features a *status variable* x that is situated in S's parent space S'. The status variable x is created when S is created and is manipulated as follows:

1. If S is merged, $x = \texttt{merged}$ is injected into S'.
2. If S becomes failed, $x = \texttt{failed}$ is injected into S'.
3. If S becomes distributable and has n alternatives, $x = \texttt{alternatives}(n)$ is injected into S'.
4. If S becomes entailed, $x = \texttt{succeeded}(\texttt{entailed})$ is injected into S'.
5. If S becomes stuck, $x = \texttt{succeeded}(\texttt{stuck})$ is injected into S'.

An additional provision, referred to as *freshening*, is needed for stable spaces that become runnable again (by application of Inject or Commit, Figure 10.4(b)):

6. If S is stable and becomes runnable again, a fresh variable y with home S' is created and S's status variable is replaced by y.

Status variable access is provided by the following primitive operation

$$\{\texttt{AskVerbose}\ x\ y\}$$

which synchronizes on x being $\xi \mapsto S$. It returns S's status variable z by pushing $y = z$.

The design of AskVerbose is simple, but suffers from a subtle possibility of hard to find programming errors. If the current space is subordinated to S, a thread can synchronize on S's status variable. Typically, this is the result of a programming error: S can never become stable due to a thread that suspends globally on z. To avoid this deadlock scenario, the application of AskVerbose is restricted to admissible spaces. For an admissible space it is guaranteed that this situation cannot occur. Section 13.3.5 clarifies that the restriction to admissible spaces is essential for the implementation.

Relation to Mozart. The Mozart implementation of spaces deviates slightly in the handling of the status variable. It uses futures (read-only variants of logic variables, see Section 3.5) instead of logic variables to offer protection against programming errors.

10.4.3 Debugging Support

A quite common situation is that a space suspends due to a programming error. Debugging tools that are programmed from spaces need to account for this situation (the Explorer is a particular example, Chapter 8). Therefore the design is extended:

7. If S becomes suspended, a fresh variable y with home S' is created and S's status variable is replaced by y. The statement $x = \mathbf{suspended}(y)$ is injected into S'.

```
fun {Deref X}
   case X of suspended(X) then {Deref X} else X end
end
fun {Ask S}
   case {Deref {AskVerbose S}} of succeeded(_) then succeeded
   [] X then X
   end
end
```

Fig. 10.5. Ask programmed from AskVerbose.

From AskVerbose it is straightforward to program Ask, as is shown in Figure 10.5.

Example 10.8. Consider the following example, where the variables X and Y are not determined and superordinated to S:

```
S={NewSpace proc {$ _} {Wait X} {Wait Y} end}
Z={AskVerbose S}
```

The thread created to execute the script for S eventually globally suspends on X. That is, the space S becomes suspended. Hence Z is determined to suspended(_). After executing X=1, the thread suspending on X resumes but globally suspends again on Y. And again S suspends, which means that Z is constrained to suspended(suspended(_)). Telling Y=1 eventually results in Z being determined to suspended(suspended(succeeded(entailed))).

10.5 Choice of Programming Language

Computation spaces presuppose the essential features of Oz Light. First-class procedures are essential for space creation and injection. Implicit synchronization is essential to synchronize on stability of spaces. Concurrency is essential for controlling speculative computations and for a clear model of attaching computations to spaces.

An additional design decision of Oz is that procedures are relational rather than functional. Relational means that results are passed as side effects on variables. This decision has no impact on the design of spaces. Any language will do, provided it offers the essential ingredients such as implicit synchronization, concurrency, and first-class procedures. Smolka describes in [136] a variant of Standard ML that offers these features. Spaces can straightforwardly build on top of this language. My paper [127] exemplifies this by using spaces for composable constraint combinators in the context of this variant of SML.

The decision to use Oz is motivated by the following facts. Firstly, as a corollary to the above discussion, the language of choice is independent of spaces. Secondly, using Oz has the advantage that all program fragments are for real. The programs can be tried with Mozart [92] as a production quality system. The programs are the abstractions that are used in Mozart. The programs serve as foundation for the thorough evaluation of the approach in this book.

11. Constraint Combinators

This chapter discusses composable concurrent constraint combinators programmed from spaces. Spaces are applied to a broad range of combinators: negation, generalized reification, disjunction, and implication (conditional). It is empirically shown that a space-based implementation of combinators is competitive with a native C++-based implementation.

11.1 Introduction

Spaces can be used to encapsulate and control speculative computations. This allows to program combinators where execution of constraints subject to combination is delegated to local spaces. The logic behind the combinator is programmed then from space operations. Whereas combinators allow to program constraints, spaces allow to program combinators. The composable setup of spaces makes space-based combinators composable to start with. Composable combinators are also known as *deep-guard combinators*.

Applications. Our experience shows that applications of constraint combinators in finite domain programming are infrequent. However, they turn out to be of great importance for other constraint domains, like feature constraints. In particular, they have turned out to be essential for computational linguistics [32], where constraints from different domains are combined naturally.

A second application area is prototyping constraints. New constraints can be developed by high level combination. After experiments have shown that they are indeed the right constraints, a more efficient implementation can be attempted. This motivation is similar to that for constraint handling rules (CHR) [37]. Spaces are primitives to combine constraints, a feature that an implementation of CHRs presupposes.

Related Work. Previous work on constraint combinators include Saraswat's concurrent constraint programming framework [117, 118], the cardinality combinator by Van Hentenryck and Deville [143], and cc(FD) [145]. The approaches have in common that the combinators considered are "flat" as opposed to "deep": the constraints that can be combined must be either built-in, or allow a simple reduction to built-in constraints (cardinality combinator).

C. Schulte: Programming Constraint Services, LNAI 2302, pp. 105–116, 2002.
© Springer-Verlag Berlin Heidelberg 2002

A deep-guard combinator has been proposed and implemented first in Nu-Prolog by Naish [96, 95]. The solution was not fully general in that reduction was limited to groundness rather than entailment. The first language with a full design and implementation of deep-guards was AKL [59, 44, 58].

The approaches mentioned so far offer a *fixed* set of combinators. Here the focus is on primitives and techniques for programming combinators. For all combinators except constructive disjunction (available in cc(FD)), it is shown how to program them from spaces.

A different approach to combining constraints are *reified* constraints (also known as metaconstraints). Reification reflects the validity of a constraint into a 0/1-variable. Constraints can then be combined by using the 0/1-variable. Spaces are not intended as a replacement for reified constraints. As is discussed in Section 11.3, a space-based reification combinator can offer better propagation in cases where reified constructions propagate poorly. Space-based reification is applicable to all expressions, including propagators for which a constraint programming system does not offer a reified version.

11.2 Concurrent Negation

This section familiarizes the reader with spaces for programming combinators by showing how to program a concurrent negation combinator from them.

For a given constraint ϕ, the negation combinator provides an implementation for the constraint $\neg\phi$. The negation combinator $\neg\phi$ executes the propagator for ϕ and:

– disappears, if the propagator for ϕ becomes failed.
– fails, if the propagator for ϕ becomes entailed.

Execution of ϕ by the negation combinator requires *encapsulation* of the computation performed by ϕ. Basic constraints that are told by propagation of ϕ must be hidden from other computations. Basic constraints that are told by other computations must be visible to ϕ. First-class computation spaces are used as encapsulation mechanism.

Some Abstractions. The following abstractions are helpful in the remainder of this chapter. Quite often no access to the root variable of a space is needed, hence it is convenient to allow a nullary procedure:

```
fun {Encapsulate P}
   {NewSpace if {ProcedureArity P}==1 then P
             else proc {$ _} {P} end
             end}
end
```

To simplify presentation the following procedure **Status** is used (for the definition of **Deref** see Figure 10.5 on page 103):

```
fun {Status S}
   case {Deref {AskVerbose S}}
   of failed             then failed
   [] succeeded(S)       then S
   [] alternatives(_) then stuck
   end
end
```

Detecting Programming Errors. A stuck space (see Section 10.4) is stable, but neither failed nor entailed. If a space S becomes stuck, it contains propagators or threads that synchronize on variables that are local to S. That is, constraint propagation within S has not been strong enough to completely drive reduction. Usually, a stuck space is the result of a programming error. In the following, this is modeled by raising an exception **error**.

The Combinator. The concurrent negation combinator takes a statement (as nullary procedure P) and creates a space running P. To make it concurrent, a new thread is created that blocks until the created space becomes stable.

```
proc {Not C}
   thread
       case {Status {Encapsulate C}}
       of failed    then skip
       [] entailed then fail
       [] stuck     then raise error end
       end
   end
end
```

11.3 Generic Reification

Reification is a powerful and natural way to combine constraints. This section presents a generic reification combinator which is shown to provide stronger propagation than constructions that use reified propagators alone.

Reification. The reification of a constraint ϕ with respect to a 0/1-variable b (a finite domain variable with domain $\{0,1\}$) is the constraint $\phi \leftrightarrow b = 1$. Whether ϕ holds is reflected into the *control variable* b as follows:

"$\Rightarrow$" If ϕ holds, $b = 1$ must hold. If $\neg\phi$ holds, $b = 0$ must hold.
"$\Leftarrow$" If $b = 1$ holds, ϕ must hold. If $b = 0$ holds, $\neg\phi$ must hold.

Having 0/1-variables b that reflect validity of constraints allows for powerful means to combine constraints. Common examples are Boolean connectives expressed by propagators (Sections 8.2 and 11.4 contain examples).

Direction "$\Rightarrow$" can be programmed along the lines of the negation combinator of Section 11.2. Suppose that S refers to a space running the statement to be reified and B refers to the 0/1-variable. Then direction "$\Rightarrow$" is as follows:

```
⟨"⇒"⟩ := case {Status S}
          of failed    then B=0
          [] entailed then B=1
          [] stuck     then raise error end
         end
```

For the case of direction "⇐" where B is determined to 0, if the space S becomes entailed, the current space must be failed. Otherwise, if S becomes failed, nothing has to be done. This behavior is already realized by the above encoding of direction "⇒".

Space Merging. Consider the case of direction "⇐" for $b = 1$. The required operational behavior includes two aspects. Firstly, a computation state must be established as if execution of σ had not been encapsulated. Secondly, if σ has not yet been completely evaluated, its further execution must perform without encapsulation.

Both aspects are dealt with by Merge. The direction "⇐" of the reification combinator is as follows:

```
⟨"⇐"⟩ := if B==1 then _={Merge S} else skip end
```

The Combinator. The reification combinator is obtained from the implementation of both directions, which must execute concurrently. Concurrent execution is achieved by creating a thread for each direction. The procedure Reify takes a procedure for the statement to be reified and returns a 0/1-variable:

```
fun {Reify E}
   S={Encapsulate E} B
in
   B::0#1 thread ⟨"⇒"⟩ end thread ⟨"⇐"⟩ end B
end
```

Example 11.1 (Relation to Propagator-Based Reification). Consider the reification of the conjunction of $x + 1 = y$ and $y + 1 = x$ with respect to the variable b, where x and y are finite domain variables. Ideally, reification should determine b to 0, since the conjunction is unsatisfiable. Posting the constraints without reification exhibits failure. To obtain a reified conjunction, the conjuncts must be reified by introducing control variables b_1 and b_2:

$$b_1 = (x + 1 = y) \land b_2 = (y + 1 = x) \land b \in \{0, 1\} \land b_1 \times b_2 = b$$

Neither b_1 nor b_2 can be determined, thus b cannot be determined.

The reification combinator developed in this section is applied as

```
B={Reify proc {$} x + 1 = y  y + 1 = x  end}
```

Both constraints are posted in the same local space S. Exactly like posting them in the toplevel space, propagation fails S. Indeed, the reification combinator determines b to 0.

This shows that using spaces for reification can yield better constraint propagation than reifying each propagator individually. Individual propagator reification encapsulates the propagation of each propagator. This in particular *disables* constraint propagation in reified conjunctions. This is a major disadvantage, since reified conjunctions occur frequently as building block in other reified constructions.

On the other hand, the generic reification combinator offers weak propagation in case the control variable is 0, because it does not impose the constraint's negation. Instead of propagation, constraints told by other propagators are tested only. Whenever a reified propagator is available, it is preferable to use it directly. So the reification-combinator offers additional expressiveness but does not replace reified propagators.

11.4 Disjunction

This section shows how to program disjunctive combinators that resolve their alternatives by propagation rather than by search.

Consider a disjunction

$$\sigma_1 \vee \cdots \vee \sigma_n$$

that is composed of n statements σ_i, where the σ_i are the disjunction's *alternatives*. A straightforward operational semantics is as follows:

1. Discard failed alternatives ($\perp \vee \sigma$ is logically equivalent to σ).
2. If a single alternative σ remains, reduce the disjunction to σ (a disjunction with a single alternative σ is equivalent to σ).
3. If all alternatives have failed, fail the current space (a disjunction with no alternatives is equivalent to $\perp$).

This operational semantics can be directly encoded by the reification operator as introduced in Section 11.3. Each alternative σ_i is reified with respect to a 0/1-variable b_i. The disjunction itself is encoded by

$$\sum_{i=1}^{n} b_i \geq 1.$$

Example 11.2 (Placing Squares). The operational semantics discussed above is driven by failure only. It can be beneficial to also take entailment of alternatives into account.

As an example consider the placement of two squares s_1 and s_2 such that they do not overlap. A well known modeling is

$$x_1 + d_1 \leq x_2 \vee x_2 + d_2 \leq x_1 \vee y_1 + d_1 \leq y_2 \vee y_2 + d_2 \leq y_1$$

The meaning of the variables x_i, y_i, and d_i is as in Figure 11.1. The squares do not overlap, if the relative position of s_1 with respect to s_2 is either left,

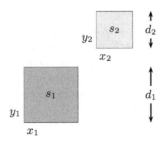

Fig. 11.1. Placing squares (Example 11.2).

right, above, or below. As soon as one of the relationships is established, the squares do not overlap.

Suppose s_1 is placed left to s_2. Since the first and second alternative are mutually exclusive (so are the third and fourth), the first and second reified propagator disappears. However, the third and fourth remain.

Assume a constraint store ϕ and a disjunction $\phi_1 \vee \phi_2$ where ϕ_1 is entailed by ϕ (that is, $\phi \to \phi_1$ is valid). Under this condition, $\phi_1 \vee \phi_2$ is logically equivalent to $\top \vee \phi_2$, which in turn is equivalent to $\top$. This justifies

4. If an alternative is entailed, reduce by discarding all alternatives.

Taking entailment into account has the advantage that execution can be more efficient, since computations that cannot contribute are discarded early. In a composable setup, this might allow for earlier reduction of other combinators and by this provide better propagation.

The implementation of the disjunctive combinator can be simplified by the following observation: it is sufficient to discard all failed alternatives but the last one. If a single alternative remains, commit to it, regardless of whether the alternative is failed or not. Merging a failed space fails the current space (see Section 10.3.2). In the following, the discussion is limited to a binary combinator. Its generalization is straightforward.

A procedure `Or` that takes two alternatives `A1` and `A2` (again encoded as first-class procedures) decomposes naturally into three parts: space creation for encapsulated execution of the alternatives, a concurrent controller, and reduction as discussed before. This yields:

```
fun {Reduce S1 S2}
    ⟨Reduction⟩
end
proc {Or A1 A2}
    S1={Encapsulate A1} S2={Encapsulate A2}
in
    ⟨Controller⟩
end
```

The controller blocks until either S1 or S2 becomes stable. This indeterminate choice is encoded by First, which returns **true** (**false**), when its first (second) argument is determined (Section 3.3.1).

```
⟨Controller⟩ := if {First thread {Status S1} end
                          thread {Status S2} end}
              then {Reduce S1 S2}
              else {Reduce S2 S1}
              end
```

The controller guarantees stability of the first space passed to Reduce.

Finally, reduction is programmed as follows:

```
⟨Reduction⟩ := case {Status S1}
             of failed    then _={Merge S2}
             [] entailed then {Kill S2}
             else case {Status S2}
                  of failed    then _={Merge S1}
                  [] entailed then {Kill S1}
                  else raise error end
                  end
             end
```

The part of Reduce that does not have a gray background executes immediately, since the controller ensures that S1 is stable. The gray part synchronizes on stability of S2. Kill kills a space by injecting failure (Example 4.2).

Reification of a statement σ can be programmed from disjunction by:

```
{Or proc {$} B=1 σ end
    proc {$} B=0 {Not proc {$} σ end} end}
```

This encoding has the disadvantage that σ is executed twice. This points out a deficiency in the designs of AKL and early versions of Oz, where neither spaces nor reification but disjunction was available as primitive.

11.5 Conditional

This section shows how to program conditionals that use arbitrary statements as conditions. In particular, it presents how to use continuations that allow to share variables between condition and body of a conditional.

A conditional consists of three constituents, all of which are statements: a *guard* G, a *body* B, and an *else-constituent* E. A suggestive syntax is

cond G **then** B **then** E **end**

The part G **then** B is called the *clause* of the conditional.

Programming a conditional from spaces is straightforward. The program used for programming Not (see Section 11.2) is adapted as follows:

```
proc {Cond G B E}
   case {Status {Encapsulate G}}
   of failed   then {E}
   [] entailed then {B}
   [] stuck    then raise error end
   end
end
```

Here G, B, and E are procedures for guard, body, and else constituent.

A common desire is to introduce variables $\bar{x}$ locally in the guard G and to subsequently use them in the body. Thus the conditional should synchronize on entailment of $\exists \bar{x} G$. In the current setup, the bindings computed for $\bar{x}$ in G are not accessible. An inefficient solution is to execute the guard again together with the body.

A more satisfactory solution is to let the guard pass the variables to the body. This can be accommodated by using the root variable of a space. In a setting with first-class procedures, sharing variables between guard and body is straightforward by letting the guard return a function for the body:

local $\bar{x}$ **in** G **fun** {$} B **end end**

Here B can refer to variables declared in the **local**-statement. Programming the conditional is now straightforward.

```
proc {Cond G E}
   S={Encapsulate G}
in case {Status S}
   of failed   then {E}
   [] entailed then B={Merge S} in {B}
   [] stuck    then raise error end
   end
end
```

Parallel Conditional. A common combinator is a *parallel* conditional that features more than a single clause with a committed choice operational semantics: As soon as the guard of a clause becomes entailed, commit the conditional to that clause (that is, continue with reduction of the clause's body). Additionally, discard all other guards.

Encoding the parallel conditional from spaces follows closely the program for disjunction in Section 11.4. In fact, the setup of the spaces for guard execution and the concurrent controller can remain unchanged.

```
⟨Reduction⟩ := if {Status S1}==entailed then
                  {Kill S2} {{Merge S1}}
               elseif {Status S2}==entailed then
                  {Kill S1} {{Merge S2}}
               else raise error end
               end
```

Adding an else-constituent is straightforward.

Clauses for Disjunction. The disjunctive combinator presented in Section 11.4 can be extended to employ clauses as alternatives. This extension is straightforward but two issues require some consideration. Firstly, when to start execution of a clause's body? Secondly, for which clause employ reduction by entailment?

Execution of the parallel conditional evaluates a clause's body B only after the clause's guard G has become entailed. This in particular ensures that the thread to compute B has terminated. A disjunctive combinator, in contrast, can already commit to a clause C if its guard G is not yet stable, provided the clause is the last remaining.

It is desirable that evaluation of C's body B starts after G has been completely executed. This is guaranteed, since procedure application synchronizes on B. And B is determined to a procedure only after full execution of the guard.

As discussed in Section 11.4, it is beneficial to consider both failure and entailment of alternatives for the disjunctive combinator. Reduction by entailment is justified by the fact that if an alternative A is entailed, it is logically equivalent to $\top$. This does apply to a clause only if its body is known to be logically equivalent to $\top$. A solution is to tag clauses as $\top$-clauses and apply reduction by entailment to $\top$-clauses only.

11.6 Andorra-Style Disjunction

The disjunctive combinator discussed in Section 11.4 resolves remaining alternatives by propagation only. On the other hand, distributors encode disjunctive information as well and are resolved by search only.

A prominent idea originating from logic programming is to combine these two aspects as follows. Reduce all disjunctive combinators as far as possible by propagation. If no further propagation is possible, choose one of the not-yet reduced disjunctive combinators and apply distribution. By this, reduction consists of interleaving deterministic (propagation) and non-deterministic (distribution) reduction, where deterministic reduction is given preference.

The idea has been conceived in the context of Prolog, where the disjunctive combinators are Horn clauses, and are, in Prolog, only reduced by search. The above described principle has been first discovered by D. H. D. Warren and has been called *Basic Andorra Principle* by Haridi and Brand in 1988 [42]. It has been independently discovered by Smolka in 1991, who referred to it as *residuation* [132]. Residuation occurred first in the work of Aït-Kaci and Nasr [6], even though there has been no explicit link to search. Similar ideas have already been explored in MU-Prolog by Naish [95].

Both styles of reduction can be combined as follows. Propagation and distribution is linked by a *control variable*. For a disjunction with n clauses, the control variable x is a finite domain variable with initial domain $\{1, \ldots, n\}$.

Failure of a clause C_i excludes i by telling $x \in \{j \mid 1 \leq j \leq n, i \neq j\}$. If x gets determined, the disjunction reduces with the x-th clause.

In addition, a distributor for the control variable x is created. This implies that if normal reduction does not suffice to reduce to a single clause, the distributor assigns a value to the control variable x and by this drives execution of the disjunction by distribution.

A different way of encoding would be to start directly from the disjunction as introduced in Section 11.4 and use the constraint $x = i$ in the guard of the disjunction. The drawback of the encoding is that it cannot be extended to handle ⊤-clauses, since the guards will not become entailed due to the constraints on the control variable.

At first sight, the demonstrated combination of propagating and distributing disjunctions looks promising. Its practical use, however, is limited due to hard-wired control. Systems built on the Andorra principle suffer from the inflexible control [116, 58]. Moolenaar and Demoen describe in [91] how selection criteria like least number of alternatives (first-fail) can be implemented on an abstract machine level. While strategies like first-fail might be appropriate in some situations, they might fall short in others. The right thing is as usual: make it programmable and provide commonly used abstractions.

Since on the other hand sophisticated selection abstractions for finite domain variables are available, the following approach works well in practice. Make the control variables for the disjunctions explicit and control distribution by distribution of the control variables.

11.7 Discussion and Evaluation

The Mozart implementation of Oz (version 1.1.0) switched from a native C++-based implementation of combinators to a space-based implementation. Information on techniques for native implementation of combinators can be found in [58, 77, 76]. The main motivation has been to simplify the implementation. The goal has been to decrease the maintenance effort which turned out to be prohibitive for the native implementation.

The native implementation requires widespread support. Support for encapsulation is needed, which is shared by the space-based implementation. The native implementation requires extension of the underlying abstract machine by several instructions and specialized data structures (instead of spaces). A particular source of problems has been the implementation of a concurrent control regime in C++ as a sequential language.

This is very much in contrast to the space-based implementation. Spaces are provided completely orthogonal, most of their supporting routines are loaded on need (for more details see Chapter 13). And spaces are concurrency-enabled to start with.

Reduction and Propagation. The runtime for composable combinators is determined by two factors: the runtime needed for reduction and the runtime for propagating constraints to spaces. The latter aspect is more important for constraint applications: typically, combinators are created once at problem setup time. Most of the computational effort is then spent on search for a solution which involves a great amount of propagation.

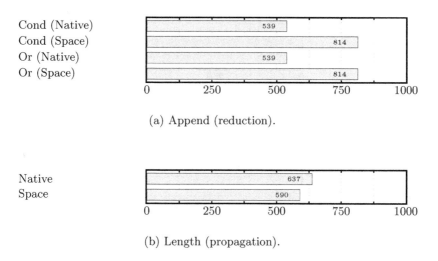

(a) Append (reduction).

(b) Length (propagation).

Fig. 11.2. Runtime (in milliseconds) for examples using combinators.

Figure 11.2(a) shows the runtime of Append. Append is the typical tail-recursive program that appends two lists with 10000 elements. Append is just concerned with reduction. Here the space-based implementation is around 50% slower than the native implementation. The reason why the native implementation is faster is that the entire reduction consists only of space and thread creation directly followed by reduction. This is faster in a native implementation since no first-class space is created and no overhead is incurred by executing Oz programs. In addition, it is clarified that both conditional and disjunctive combinator have the same runtime behavior.

Figure 11.2(b) shows the runtime for the example Length. Length computes the length of a list, where a disjunctive combinator is used to propagate the length of a list to and from a finite domain variable. This is iterated a hundred times as follows: from a list the length is computed, from the length a list is computed, ..., and so on. The runtime for Length is thus dominated by propagation. Here the space-based implementation is even slightly more efficient than the native implementation. The reason is that due to the simplifications that have become possible by removing the native implementation of combinators, the entire implementation has become more efficient.

These experiments suggest that the space-based implementation is competitive to the native C++-implementation as it comes to runtime. Native combinators are slightly faster for programs where execution time is dominated by reduction of combinators. For examples where the runtime is dominated by constraint propagation, both approaches offer approximately the same execution speed.

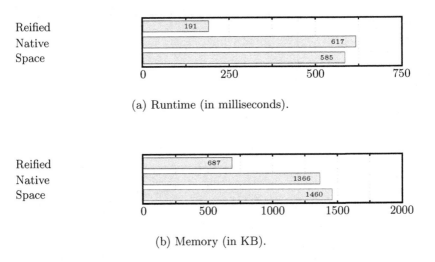

(a) Runtime (in milliseconds).

(b) Memory (in KB).

Fig. 11.3. Runtime and memory requirements for implementations of Bridge.

Comparison to Reification. Figure 11.3 shows runtime and memory requirements for the Bridge example. Bridge is a small scheduling problem (Appendix A.1). A central constraint for Bridge is that two tasks that require the same resource do not overlap in time. The formulation used is the single-dimensional variant of the non-overlap constraint for squares as discussed in Example 11.2. Here the runtime is dominated by propagation which explains why native and space-based implementation offer the same runtime. They also require roughly the same memory. The comparison to an implementation that uses propagator-based reification stresses that combinators are not a replacement for reified propagators but rather an addition.

Figure 11.3(b) reveals that space consumption of space-based and native combinators is approximately the same.

12. Implementing Oz Light

This chapter outlines the implementation architecture of Oz Light. The architecture serves as foundation for the implementation of spaces.

12.1 Overview

The chapter is concerned with the implementation aspects of Oz Light that are fundamental for the implementation of spaces. The first issue is the store and in particular the variables in the store. The second issue is synchronization and control of threads.

The implementation is sequential: there is at most one thread executing at a time. The architecture features the following components:

Store. The store implements the constraint-graph. Its nodes are the variables and the values. Its edges represent equality-constraints between nodes. The central operation on the store is tell which possibly inserts new edges into the graph.

Emulator. The emulator executes threads. Execution possibly creates new threads, creates new nodes in the store, and performs tell operations on the store. The details of the emulator are of little concern for spaces. It suffices, that after execution of a thread, its status and its suspension set is available from the emulator.

Scheduler. The scheduler is the implementation's main control instance. It maintains a pool of runnable threads and provides fair execution control for runnable threads. The scheduler creates, suspends, wakes, and terminates threads.

How the actual statements of Oz Light are implemented is orthogonal to the implementation of spaces. A sketch of a complete implementation of a language similar to Oz Light is [77]. Scheidhauer discusses implementation and evaluation of the emulator in [119]. Mehl discusses the implementation of store and scheduler in [76].

C. Schulte: Programming Constraint Services, LNAI 2302, pp. 117–120, 2002.

12.2 Synchronization

The implementation's most distinguished service is thread synchronization. When a thread T suspends, its topmost statement is in charge of computing the suspension set $\mathcal{S}(T)$. As soon as new constraints on a variable in $\mathcal{S}(T)$ become available, T is woken. To implement synchronization efficiently, T is attached to all variables in $\mathcal{S}(T)$.

As soon as new constraints become available on a variable x, the attached threads are made runnable by waking them. The set of suspended threads $\mathcal{S}(x)$ attached to a variable x is called the *suspension set of x*. Waking the threads in $\mathcal{S}(x)$ is also called to *wake the variable x*.

The emulator provides access to the suspension set of thread T. The scheduler attaches a suspended thread T to the variables $\mathcal{S}(T)$. The store detects which variables must be woken. Again, the scheduler wakes the variables.

A thread that is woken is *not guaranteed* to make any progress. Even though new constraints are available, the topmost statement might still be unable to reduce. For example, if a thread synchronizes on $x + y = z$ and only x has become determined while y is still unconstrained. The reason is that the implementation does not track how much information is needed to resume a thread. It just makes the safe but pessimistic approximation that new constraints on a variable wake the thread.

12.3 Store

The constraint store implements the *constraint-graph*. Its nodes represent variables and values. Its edges represent equality-constraints between variables and values.

Nodes. The constraint store has three different kinds of nodes (Figure 12.1). A *variable node* has a link to its suspension set. A *value node* for a simple value has no outgoing edges. A value node for a tuple node f/n has n outgoing edges that point to the nodes for its n subtrees. A *reference node* is created by constraining variables. It has a single outgoing edge pointing to an arbitrary node. In the following, nodes and their entities are identified.

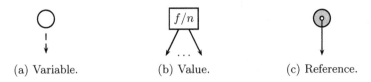

(a) Variable.   (b) Value.   (c) Reference.

Fig. 12.1. Nodes in the constraint store.

The implementation combines all store compartments in that it has nodes for procedures and ports. For example, a procedure node has two links to implement a closure: to the free variables and to the code (statement).

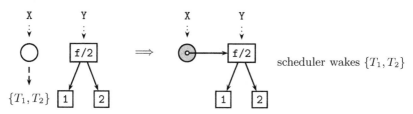

Fig. 12.2. Binding a variable node.

Binding Variables. Ideally, telling $x = v$ would redirect all links that point to x to v instead. This is not feasible: there is no simple way to efficiently maintain the incoming edges of x. Instead, a variable node is turned into a reference node. Figure 12.2 shows how a new constraint is added to the store. X refers to a variable node, whereas Y refers to a value node. Telling X=Y to the store wakes X and turns the variable node into a reference node pointing to Y. Turning a variable node into a reference node is called to *bind the variable (node)*. If a variable node x is bound to another variable node y, both variables are woken: new information is available on both x and y.

Reference nodes in the store are transparent. Routines that access the store implicitly follow links from reference nodes ("dereferencing"). A garbage collector, for example, is free to remove reference nodes ("path compression").

Unification. Unification is used to achieve equality between two subgraphs in the store. An example is sketched in Figure 12.3. As far as spaces are concerned, unification adds edges into the store and possibly wakes variables. For more details of unification in the context of Oz, consider Mehl's thesis [76].

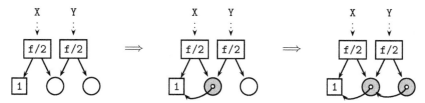

Fig. 12.3. Unification example.

12.4 Scheduler

The scheduler is the implementation's main control instance. It maintains a pool of runnable threads and provides fair execution control to the pool. It controls thread transitions: the scheduler creates, suspends, wakes, and terminates threads.

Thread Selection. The scheduler maintains the *runnable pool* containing runnable threads. When a new thread is created, it is added to the runnable pool. When a thread is woken, it is also added to the runnable pool.

The scheduler selects threads fairly. This is typically achieved by maintaining the runnable pool as a queue. The thread selected for execution is called the *current thread.*

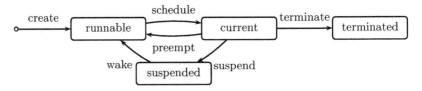

Fig. 12.4. Thread states and their transitions.

Thread State. Figure 12.4 shows thread states and the transitions between them. The scheduler controls execution of runnable threads. After selection of the current thread T, the scheduler applies the emulator to T. After execution of T stops, the scheduler takes the necessary actions depending on T's execution status:

Terminated. The thread T has been completely executed and is discarded.

Preempted. The thread T is still runnable but has used up its time slice. It is entered into the runnable pool to be run again later. Preemption together with organizing the runnable pool as queue guarantees fairness.

Suspended. The topmost statement σ of T has blocked and cannot reduce. The statement σ itself decides on which variables it synchronizes. The suspension set $\mathcal{S}(T)$ is available from the emulator. The scheduler enters T to all variables in $\mathcal{S}(T)$.

Runnable Threads. Since a thread can suspend on more than a single variable, threads contained in a variable's suspension set can already be runnable. Therefore, a thread carries a mark that identifies it as being runnable. A thread is entered into the runnable pool only if it is not yet marked as runnable. As a corollary, a thread contained in a suspension set of a variable can also be terminated. In the following, it is assumed that the scheduler takes care of runnable and terminated threads during waking.

13. Implementing Spaces

This chapter discusses the implementation of computation spaces. The implementation of Oz Light is extended by nodes for spaces and situated entities, by a scheduler that handles situated threads and tests stability, and by the space operations proper.

13.1 Overview

The central points in implementing first-class computation spaces are:

Space Tree (Section 13.2). The space tree is implemented by space reference nodes which provide first-class access to space nodes. The implementation operates on space nodes for all but first-class access. Situated nodes implement links to home spaces as required for situated entities. Multiple stores are simulated by a single store that provides the view to the constraints for a single space at a time.

Stability (Section 13.3). The implementation of stability covers two aspects. The first aspect is the information required to detect stability and how the information is maintained as computation proceeds. The second aspect is how and when to actually test a space for stability.

Merge (Section 13.4). Merging is an involved transformation of the space tree, since it simultaneously changes the home of a large number and variety of data structures. In particular, merging must consistently maintain stability information.

Search (Section 13.5). The implementation of search is concerned with distributors and cloning. While cloning resembles many aspects of copy-based garbage collection, it also features unique aspects.

Richer Basic Constraints (Section 13.6). Stores are extended to cover aliasing of variables, tree constraints, and finite-domain constraints.

Ports (Section 13.7). Message-sending across space boundaries and in particular sendability is discussed.

Performance Overview (Section 13.8). The last section gives an overview of the performance of space operations.

C. Schulte: Programming Constraint Services, LNAI 2302, pp. 121–141, 2002.

13.2 Space Tree

This section is concerned with data structures for spaces and extensions needed for multiple stores and situated computations.

13.2.1 Nodes and Links

The model for first-class computation spaces (Chapters 4 and 10) separates the first-class reference to a space (a name) from the space proper. The implementation follows this setup. First-class references are implemented by *space reference nodes*. Spaces proper are implemented by *space nodes*.

Space Reference Nodes. Space reference nodes introduce a new type of node to the store. A space reference features a link that points to a space node. The only purpose of a space reference is to provide first-class reference to space nodes.

Space Nodes. A space node is implemented by a data structure that organizes the space nodes as a tree ("parent link"). The implementation of spaces is mostly concerned with space nodes. This justifies that space abbreviates space node in the following. In addition to the parent link, a space node has links to all components of a space: root variable, status variable, and so on.

Current Space Node. The implementation maintains the *current space*. The current space is set by the scheduler: when a thread T is selected as current thread, the current space is set to $\mathcal{H}(T)$. Making a space S current *installs the space*, which in particular involves the installation of the store of S (to be discussed later).

Situated Nodes. Threads, variables, procedures, and ports are situated in their home space. This is implemented by a link pointing to the appropriate space node ("home link"). Upon creation, the home link is initialized to point to the current space node.

Failed Nodes. A failed space node carries a mark that identifies it as failed.

Discarded Nodes. A space S_1 is discarded *implicitly*, if a space $S_2 \in \uparrow S_1$ fails, since S_2 has no access to S_1. Hence, discarded nodes require an explicit test. The test traverses the space tree by following parent links from S_1 until a failed node or the toplevel node is encountered. In case a failed node is encountered, S_1 is discarded.

Safe and Unsafe Links. Due to different directions of links with respect to the space tree, links can be either safe or unsafe.

Upward Links. Home links and parent links point *upwards* and are *safe*. While computation proceeds in the current space, upward links are guaranteed to refer to non-failed spaces.

Downward Links. Links to threads and to spaces as implemented by space references point *downwards* and are *unsafe*. Links to threads are stored in suspension sets of variables and in the scheduler's runnable pool. Downward links can point to failed or discarded spaces. An attempt to follow an unsafe link must always be preceded by a test whether the referred space is failed or discarded.

Garbage Collection. An obvious advantage of separating space references from space nodes is a factorization of concerns. An additional advantage is that the potential for memory reclamation during garbage collection is increased. Failed space nodes (as well as merged space nodes) are not retained during garbage collection. If the space reference remains accessible, it is marked appropriately such that space operations can test whether the operation is applied to a failed or merged space. This is sufficient, since no operation needs access to failed and merged space nodes.

13.2.2 Threads

Failure introduces new states and new state transitions for threads (Figure 13.1). When a space S is failed, all threads in $\Downarrow S$ are discarded. Both runnable and suspended threads can be discarded. Discarding is implicit as opposed to the other state transitions that are performed by the scheduler. An additional transition is that suspended threads can be created (discussed later).

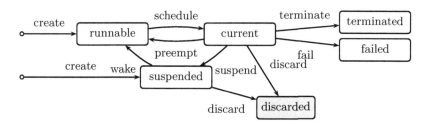

Fig. 13.1. Thread states and their transitions including failure.

Waking Threads. Waking takes into account that threads are situated and can be possibly discarded. Execution of a tell statement $x = n$ in space S wakes only threads T with $\mathcal{H}(T) \in \Downarrow S$. The test whether $\mathcal{H}(T) \in \Downarrow S$ traverses the space tree starting from $\mathcal{H}(T)$ until either S or the toplevel is encountered. If S is encountered, $\mathcal{H}(T) \in \Downarrow S$ and T is woken. Additionally, the test reveals whether T is discarded, in which case T is dropped from x's suspension set.

The Scheduler. The scheduler as main control instance of the implementation is enhanced in order to support computation spaces. It maintains the current space, which refers to the home space of the current thread. The scheduler works as follows:

1. Select a runnable thread T from the runnable pool.
2. If T is discarded, continue with 1.
3. Install $\mathcal{H}(T)$. If installation fails, continue with 1. Otherwise, T becomes the current thread and $\mathcal{H}(T)$ becomes the current space.
4. Run T.
5. Test $\mathcal{H}(T)$ for stability (Section 13.3).
6. Continue with 1.

The test whether T is discarded is necessary. The test during waking is not sufficient, since T might have been discarded after waking.

13.2.3 The Store: Model

The model introduced in Chapter 10 defines that each space has a private constraint store. The private store is inherited from the parent upon space creation and a tell in S is repeated in all spaces in $\downarrow S$. The implementation improves over the naive model in that it maintains a single store shared among all spaces. The single store facilitates sharing of common constraints, avoids repeated tells, and conservatively extends the store implementation of Oz Light.

The implementation of the single store is introduced in two steps. Firstly, an abstract model of a single store together with the invariants that make it faithful with respect to the naive model is introduced. Secondly, a concrete implementation based on the abstract model known as scripting is described.

In the sequel, a simplified version of the store is described that contains only basic constraints $x = n$. Extensions are considered in Section 13.6.

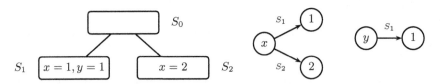

Fig. 13.2. Example space tree and corresponding full graph.

The Full Graph. The store is modeled by a single graph, referred to as the *full graph*. Edges in the full graph point from a variable node x to a value node n and are labelled by a space S. An edge in the full graph is referred to by $x \overset{S}{\to} n$. The key point is that a variable node can have multiple outgoing edges labelled by different spaces. An example space tree together with its corresponding full graph is shown in Figure 13.2 (the home of both x and y is S_0).

The full graph maintains the following invariants:

Situatedness.

$$x \overset{S}{\to} n \quad \Longrightarrow \quad \mathcal{H}(x) \leq S$$

This is obvious: a variable x is only visible in spaces $\Downarrow \mathcal{H}(x)$.

Orthogonality.

$$x \overset{S}{\to} n \quad \text{and} \quad x \overset{S'}{\to} n' \quad \Longrightarrow \quad S \not\leq S' \quad \text{and} \quad S' \not\leq S$$

This invariant guarantees both consistency and minimality.

Consistency. If $x \overset{S}{\to} n$, then for all $S' \in \downarrow S$ there is no $x \overset{S'}{\to} n'$ with $n \neq n'$.

Minimality. If $x \overset{S}{\to} n$, then for all $S' \in \downarrow S$ there is no $x \overset{S'}{\to} n$.

If $x \overset{S}{\to} n$ and $\mathcal{H}(x) = S$, both invariants together guarantee that there is no other link for x. This in particular entails that a link $x \overset{S}{\to} n$ for the toplevel space S is the only link for n.

The single operation on the graph is an attempt to tell $x = n$ in a space S. A failed tell attempt fails S and removes all edges $x' \overset{S'}{\to} n'$ for all x', n', and all $S' \in \Downarrow S$. Execution of the tell covers the following cases:

Equal Below. $x \overset{S'}{\to} n$ and $S < S'$: Remove $x \overset{S'}{\to} n'$ and insert $x \overset{S}{\to} n$ (minimality).

Different Below. $x \overset{S'}{\to} n'$, $S < S'$, and $n \neq n'$: Fail S' and insert $x \overset{S}{\to} n$ (consistency).

Equal Above. $x \overset{S'}{\to} n$ and $S' \leq S$: Do nothing (minimality).

Different Above. $x \overset{S'}{\to} n'$, $S' \leq S$, and $n \neq n'$: Fail S (consistency).

New. Otherwise, insert $x \overset{S}{\to} n$.

Example 13.1 (Operations on the Full Graph). Consider the following tell attempts to the full graph shown in Figure 13.2.

- Tell $y = 2$ in S_2: "New" applies (Figure 13.3(a)).
- Tell $x = 1$ in S_0: "Equal Below" for S_1 and "Different Below" for S_2 applies (Figure 13.3(b)).
- Tell $y = 1$ in S_1: "Equal Above" applies (space tree and full graph remain unchanged).

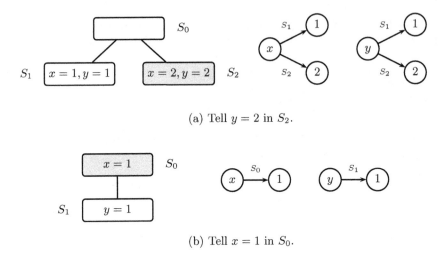

(a) Tell $y = 2$ in S_2.

(b) Tell $x = 1$ in S_0.

Fig. 13.3. Results of tells in Example 13.1.

13.2.4 The Store: Implementation

Scripting realizes a subgraph of the full graph that corresponds to the current space S and supports switching to a different space S'. The subgraph contains all links for spaces in $\Uparrow S$. If S is current, all spaces in $\Uparrow S$ are said to be *installed*. The subgraph is referred to as *installed graph*. Due to the orthogonality invariant, the installed graph has the property that a variable node can have at most one outgoing edge. This allows to conservatively extend the implementation of Oz Light to handle multiple bindings in different spaces.

When switching the current space from S_1 to S_2, constraints (links) for S_1 must be deinstalled while the constraints (links) for S_2 must be installed. Each space maintains a *trail* and a *script*, which are as follows.

– If S is installed, the script is empty. Otherwise, the script records all links $x \overset{S}{\to} n$ with $\mathcal{H}(x) < S$.
– If S is installed, the trail records all variables x with $x \overset{S}{\to} n$ and $\mathcal{H}(x) < S$. Otherwise, the trail is empty.

Both script and trail care about speculative constraints only, that is about links $x \overset{S}{\to} n$ with $\mathcal{H}(x) < S$. Suppose that S' is the current space and $\mathcal{H}(x) = S$. Either the link is installed anyway ($S' \in \Uparrow S$), or it cannot be observed ($S' \notin \Uparrow S$). As a consequence, both trail and script for the toplevel space are empty.

Supervisor Threads. Space switching is thread-driven. When the scheduler picks a thread T for execution, $\mathcal{H}(T)$ is made current by installation. This means that installation of a space S can be requested by creating a thread in S.

The implementation of scripting requires that the constraints contained in scripts are supervised by *supervisor threads.* Therefore the implementation maintains the following invariant. If the script of S is not empty, then S is either runnable or for each variable x in the script, there exists a supervisor thread T with $\mathcal{H}(T) = S$ that suspends on x.

Supervisor threads are very attractive for detecting stability in Section 13.3. This is due to the fact that if a non-installed space has speculative constraints, it also has a speculative thread. Hence, considering threads for stability is sufficient.

Single Supervisor Thread. At most one supervisor thread for all variables in the trail is sufficient. Mehl [76] observes, that if S is runnable, no supervisor thread needs to be created, since S is installed eventually. This can be even further optimized. In case there is already a thread in S that synchronizes on x, no supervisor thread for x is needed.

Variable Binding. Binding a variable node x in S checks whether the binding is speculative $(\mathcal{H}(x) \neq S)$. If the binding is speculative, x is recorded on the trail before turning it into a reference node.

Binding a variable covers all cases as for telling to the full graph. "Equal Above", "Different Above", and "New" are as in Oz Light, since they deal with links that are currently installed. "Equal Below" and "Different Below" are handled by supervisor threads. If the variable node x is bound in space S and there is a constraint for x in the script of space $S' \in {\downarrow}S$, then there exists a supervisor thread T that suspends on x with $\mathcal{H}(T) = S'$. When binding x in S, T is woken and eventually S becomes installed.

Script Deinstallation. A new supervisor thread T is created. For each variable node x stored in S's trail the pair $\langle x, n \rangle$ is put in the script of S. Here n is the value to which x is constrained. Simultaneously, the variable node for x is reestablished and T is added to x's suspension set. After deinstallation the parent of S is installed.

Script Installation. A space is installed by telling $x = n$ for all script entries $\langle x, n \rangle$. As a consequence, installation can fail due to a failed tell attempt. If installation has been requested by a supervisor thread, cases "Equal Below" and "Different Below" are handled by installation.

No threads need to be woken during installation. If the script of S contains $\langle x, n \rangle$, the constraint $x = n$ has been available when S has previously been installed. Hence, a thread that suspends on x must have been woken while S has been installed. The situation is slightly different for supervisor threads since they are created during deinstallation. But their very purpose is that they are woken by tells in superordinated spaces only.

Space Switching. Switching between arbitrary spaces S_1 and S_2 iterates single step installation and deinstallation. All spaces up to the toplevel space are deinstalled and all spaces from the toplevel space to S_2 are installed. This can be optimized by deinstalling only up to the closest common ancestor space of S_1 and S_2.

Example 13.2 (Scripting). Let us consider as an example for scripting the situation as displayed in Figure 13.2. The corresponding store and space tree with S_0 as current space is shown in Figure 13.4(a). Installed space nodes have a gray background. Space nodes which are not installed are displayed with their script as content. The nodes for the store are as introduced in Section 12.3. The threads T_i are the supervisor threads for S_i.

Telling $x = 1$ in S_0 executes as follows (shown in Figures 13.4(b) to 13.4(e)):

– The tell binds x and wakes the threads T_1 and T_2.
– Running T_2 fails S_2.
– Running T_1 installs S_1. The binding of y is recorded on the trail of S_1.
– Deinstallation of S_1 creates a script entry $\langle y, 1 \rangle$ and a new supervisor thread T_3.

Single Trail. Each space has a private trail. This can be optimized by using a single trail common to all spaces. The single trail has multiple sections separated by marks. Each section corresponds to an installed space. The topmost section corresponds to the current space. Entries are made only for the current space, that is, to the topmost section.

Tells in Arbitrary Spaces. Later the need arises to perform tells $x = v$ in a space S different from the current space (merging is a particular example, Section 13.4). This can be accommodated by creating a thread in S with $x = v$ as its single statement.

Related Work. Supporting multiple variable bindings simultaneously depending on the computational context occurred first in the area of Or-parallel Prolog implementations [19]. Gupta and Jayaraman classify in [39] approaches according to the cost of three essential operations: creation (space creation, here), switching (space switching, here), and binding-lookup. It is argued that at most two operations can be implemented in constant time. A formal proof of this fact can be found in [112]. Scripting decides to make creation and lookup constant time.

Scripting has the following advantages. It is a conservative technique that allows to stick to a single outgoing link per variable. Computations in a space do not pay any overhead for looking up bindings. Enforcing consistency and minimality comes for free by using threads. Additionally, supervisor threads are convenient for detecting stability.

The main disadvantage of scripting is that it is an inherently sequential technique because it only supports the view for a single space at a time. The

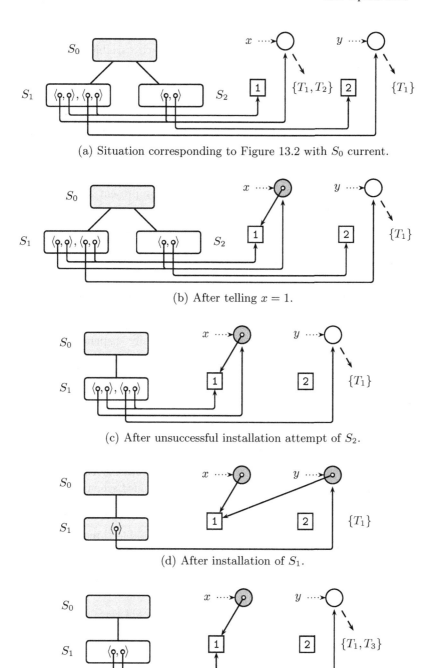

(a) Situation corresponding to Figure 13.2 with S_0 current.

(b) After telling $x = 1$.

(c) After unsuccessful installation attempt of S_2.

(d) After installation of S_1.

(e) After deinstallation of S_1.

Fig. 13.4. Computation states for Example 13.2.

Penny system, a parallel implementation of AKL, uses a different solution that provides multiple views at a time [90, 89]. Each variable maintains a list of speculative bindings indexed by the space for which the binding is valid. Montelius shows in [89] that this solution is efficient, since speculative bindings are infrequent.

Podelski and Smolka study situated simplification in [102] as a technique for detecting entailment and failure of rational tree constraints used in local computation spaces. In particular, the presented techniques are proven correct. Smolka and Treinen discuss scripting for testing entailment of record constraints [137]. Scripting is not fully incremental in that script installation and deinstallation redo work. Podelski and Van Roy present a truly incremental algorithm in [103].

13.3 Stability

The implementation of stability deals with two aspects: maintaining information on runnable and suspended threads, and testing stability based on that information. To decide whether a space S is stable, it is necessary to know whether $\Downarrow S$ contains runnable threads or globally suspended threads. To decide whether a stable space S is stuck, it is necessary to know whether S contains locally suspended threads. Semi-stability and distributable threads are discussed in Section 13.5.1.

13.3.1 Runnable Threads

Each space S maintains a *runnable counter* r_S. The runnable counter r_S is zero, if and only if $\Downarrow S$ contains no runnable thread. Rather than counting all threads in $\Downarrow S$, the implementation employs a *cascaded* scheme as follows: r_S counts the number of runnable threads in S plus the number of runnable children of S.

The runnable counter is incremented when a runnable thread is created and when a suspended thread is woken. Incrementing r_S is done as follows: r_S is incremented by one and if r_S has been zero before, incrementing continues with the parent of S. Incrementing possibly continues until the toplevel space is reached.

The runnable counter is decremented when a thread terminates or suspends. Decrementing is dual to incrementing: if, after decrementing, the value of r_S is zero, decrementing continues with the parent of S.

If a space S becomes failed, all runnable threads in $\Downarrow S$ are discarded simultaneously by decrementing the runnable counter of S's parent. Runnable threads that now have become discarded are still in the runnable pool but the scheduler can safely ignore them.

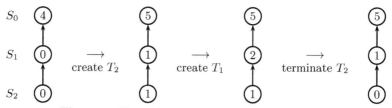

Fig. 13.5. Example for managing runnable counters.

Example 13.3 (Managing Runnable Counters). Suppose the space tree as sketched to the left of Figure 13.5. Creation and termination of threads T_1 and T_2 with $\mathcal{H}(T_i) = S_i$ results in values for the runnable counters as shown in the subsequent space trees.

A different and naive design would be to maintain the number of runnable threads in $\Downarrow S$ for each space S. The disadvantage compared to the cascaded counting scheme is obvious: if a thread in S becomes runnable (suspended), the numbers in all spaces in $\Uparrow S$ must be incremented (decremented). The cascaded scheme stops incrementing (decrementing) as soon as the first runnable space is encountered while traversing $\Uparrow S$ upwards. This in particular entails that creating and waking a thread T where $\mathcal{H}(T)$ is already runnable does not require any traversal of the space tree.

An additional advantage of the cascaded scheme is that it supports failure well. All runnable threads in a subtree can be discarded without requiring explicit access. This in particular allows to consider discarded threads in the runnable pool as garbage during garbage collection. Garbage collection of discarded threads together with the non-cascaded counting scheme has been a constant source of problems in early implementations of Oz.

13.3.2 Globally Suspended Threads

A blocked space S is stable, if the spaces in $\Downarrow S$ do not contain threads which globally suspend for spaces in $\Uparrow S$ (speculative constraints are discussed later). Each space S maintains the threads in $\Downarrow S$ that globally suspend for a space in $\Uparrow S$. Unfortunately, a scheme similar to the runnable counter is not sufficient. This is due to the fact that if a thread globally suspends for S, it does not necessarily globally suspend for spaces in $\Uparrow S$.

Hence, the implementation maintains a *global suspension set* for each space S that contains threads that globally suspend for spaces in $\Uparrow S$. As with suspension sets of variables, the implementation allows inclusion of already discarded threads. When testing whether a space has globally suspended threads, discarded threads are removed.

When a thread T globally suspends for S, the thread is inserted into the global suspension sets of all spaces in $\Uparrow\!\mathcal{H}(T) - \Uparrow S$. Insertion is performed by traversing the space tree starting from $\mathcal{H}(T)$ up to S.

Waking a thread T removes it from all global suspension sets in $\Uparrow \mathcal{H}(T)$. This is optimized by marking globally suspended threads upon suspension. Only if a thread carries a global suspension mark, removal from global suspension sets is considered.

13.3.3 Speculative Constraints

An important insight is that testing for speculative constraints comes for free. This is a consequence of supervisor threads: if S_2 contains a speculative constraint for S_1, it is guaranteed that there is a supervisor thread situated in S_2 that globally suspends for S_1.

Only testing stability of installed spaces (in particular for the current space) needs special attention. Here the supervisor thread has not yet been created. But the needed information is already available: a space has speculative constraints if its trail is not empty.

13.3.4 Local Threads

A succeeded space S is stuck, if it contains threads. The number of threads situated in S is maintained for each space S. The number is incremented upon thread creation and decremented upon thread termination.

13.3.5 Checking Stability

When a space S becomes stable, the scheduler makes the stability information available by telling the status to S's status variable. The information is told as soon as a space becomes blocked (**AskVerbose** in Section 10.4.3). Hence, the scheduler is concerned with suspended and stable spaces.

The stability test is performed directly after updating the stability information:

1. If the current space S_C is runnable, continue with the next thread.
2. Deinstall S_C (which is blocked) and make its parent current.
3. Tell S_C's status to the status variable of S_C. Tells to S_C's status variable are possible, since the current space (S_C's parent) is the home of S_C's status variable.
4. Check spaces in $\Uparrow S_C$ by traversing the space tree starting from the current space upwards, until the first blocked space S_b is encountered. Inject a new thread T running **skip** into S_b: eventually S_b is checked on behalf of T.

Bottom-Up Checking and Admissibility are Essential. Stability checking is performed bottom-up. This is necessary due to tells on status variables. A tell on the status variable for S possibly wakes threads in $\mathcal{A}(S)$ (the set of admissible spaces with respect to S is introduced in Section 10.3.2). Hence spaces in $\mathcal{A}(S)$ can become runnable. Proceeding bottom-up together with admissibility guarantees that checking S cannot make S runnable by waking threads in $\downarrow S$.

Stability Checking Is Complete. If S becomes stable, S is eventually checked for stability. This is obviously the case for blocked spaces: the thread of S that terminates last, lets the scheduler check S. Suppose that S is globally suspended and S's global suspension set contains T. Let us first consider the case $\mathcal{H}(T) = S$. Then S can become stable only if T becomes runnable first (failure is trivial). Then, S is installed and thus checked for stability. If $S < \mathcal{H}(T)$, the thread T is contained in $\mathcal{H}(T)$'s global suspension set. This means that S is eventually checked on behalf of a thread injected to S.

13.4 Merge

The implementation of space creation and injection is straightforward given the material presented in the previous sections. In contrast, merging performs a possibly involved transformation of the space tree.

When a space S_1 is merged to a space S_2, S_1 is called the *source* and S_2 the *destination* of the merge operation. The destination of a merge operation is always the current space.

Making Space Nodes Transparent. All entities situated in the source S_1 must become situated in the destination S_2. All entities situated in S_1 carry a link to S_1. The implementation uses the same technique as for binding variables: S_1 is marked as merged, and a reference to S_2 is stored in S_1. Subsequent accesses consider a merged space transparent and follow the reference until an unmarked space is encountered ("dereferencing"). Also the access to a space's parent uses dereferencing.

Testing Admissibility. The source S_1 must be admissible for the current space S_2. Admissibility ($S_1 \notin {\Uparrow} S_2$) is checked by traversing ${\Uparrow} S_2$. If, while traversing, S_1 is encountered, S_1 is not admissible. If the toplevel space is reached, S_1 is admissible.

Speculative Constraints. The speculative constraints of the source S_1 are made available to S_2 by installing the script of S_1. Since script installation possibly tells constraints new to S_1, threads are woken. Note that script installation can fail S_2.

Runnable Threads. To better understand how the runnable counter of the destination S_2 is updated, upward and downward merges are discussed separately (Figure 13.6).

For an *upward merge* (Figure 13.6(a)), the destination S_2 is the parent of the source S_1. Suppose that S_1 is runnable and has n runnable child spaces and threads. S_2 looses the runnable child S_1. Hence, its runnable counter is

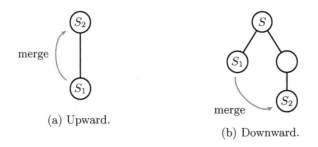

(a) Upward.

(b) Downward.

Fig. 13.6. Space trees for downward and upward merge.

decremented by one. On the other hand, its runnable counter is incremented by n, since it acquires n runnable threads and spaces. The runnable counters in $\uparrow S_2$ remain unchanged, since S_2 remains runnable (the current thread executes the merge in S_2). In case S_1 is blocked, nothing is done.

For a *downward merge* (Figure 13.6(b)), the source's parent S is an element of $\uparrow S_2$. Again, if S_1 is blocked, nothing is done. Suppose that S_1 is runnable. Then S looses a runnable child, hence its runnable counter is decremented by one. And again, the runnable counter of S_1 is incremented by the value of S_2's runnable counter. Upward and downward merge have the same effect, the distinction is to ease explanation.

Upward merges are common: merging a solution space computed by a search engine is upward. Example 10.3 discusses the use of downward merges for partial evaluation.

Globally Suspended Threads. Let us first consider an upward merge. If a thread T in S_2 is globally suspended for a space in $\Uparrow S_1$, T is already contained in the global suspension sets of $\Uparrow S_1$.

A downward merge is more involved. Consider a globally suspended thread T in S_2. The thread T is at least globally suspended for S (the parent of S_2). Hence, after merging, T becomes globally suspended for S' with $S \leq S' \leq S_1$. Thus T is entered into the global suspension sets of all spaces S'.

Installation of S_2's script during merging possibly changes the global suspension set of S_2. However, the order of script installation and global suspension set update is insignificant. If a thread T is woken by script installation, it will be removed from all global suspension sets anyway.

Local Threads. The number of local threads of S_2 is incremented by the number of local threads of S_1.

The Operation. The merge operation {Merge x y} is as follows:

1. Test whether x is determined to a space reference. If not, suspend on x or raise an exception.
2. Access the space node S_1 from the space reference node.
3. Raise an exception, if S_1 is merged or not admissible.

4. Raise an exception, if both S_1 and S_2 are distributable.
5. Fail the current space, if S_1 is failed.
6. Merge the space node S_1 with the current space node S_2:
   a) Draw a link from S_1 to S_2.
   b) Install the script of S_1 and possibly mark S_2 as failed.
   c) Incorporate stability information of S_1.
7. Mark S_1 as merged.
8. Tell that S_1 is merged (inject a thread, if merge is downward).
9. Constrain the root variable of S_1 to y.

13.5 Search

13.5.1 Choose and Commit

The implementation provides more general support for distributors than actually required by Choose. Choose itself is obtained from this more general support. This more general support allows distributors to be written in C++. For example, the Mozart implementation of Oz implements standard distribution for finite domain variables in C++.

A distributable space node contains a reference to a distributor. A distributor provides support for creation, it can be queried for its numbers of alternatives, and it provides functionality to commit to an alternative.

Distributor Creation. When a new distributor is created, it is passed a newly created variable x serving as synchronization variable. The thread that has created the distributor immediately suspends on x. A reference to x and the number of alternatives is stored by the distributor. If the space is already distributable, an exception is raised.

Semi-stability. The stability test checks whether a stable space is distributable. If the space is not distributable, execution proceeds as before. If the space has a unary distributor D, the synchronization variable of D is determined and execution of the suspending thread can proceed. Otherwise, the status variable is determined to the tuple alternatives(n), where n is the number of alternatives.

Commit. Invocation of the commit operation, as provided by the distributor D returns the number k of alternatives D has after performing the commit. If k is zero or one, D is discarded. If k is one, D has at least determined the synchronization variable. In general, D has performed a tell to the constraint store or has created a new thread. Additionally, a fresh status variable is created that is bound to alternatives(k).

13.5.2 Cloning Spaces

Cloning makes a copy of a space S. All objects that are reachable from S are copied by graph copying that preserves sharing. Recursively copying a graph of objects is well-known from copying garbage collectors (for example [155, 61]). The following aspects need to be taken into account for cloning spaces.

Stability is Essential. Copying assumes that no links point into the space S to be copied. If there are links, these links must be copied as well. This would imply a traversal of *all* data structures rather than only those reachable from S. Stability ensures that there are no links that point into S. The only links that can point downward (Section 13.2.1) are links to runnable or globally suspended threads. A stable space does neither contain runnable nor globally suspended threads.

Retaining the Original. During garbage collection, the original is discarded. Graph copying changes the original objects by marking and storing forward pointers. Thus the original space must be restored after copying. The implementation uses a trail for the information needed to reestablish the original.

Taking Situatedness into Account. A stable space S typically contains references to entities that are situated in $\uparrow S$. Entities situated in $\uparrow S$ must not be copied. Suppose that S_1 is copied and that the home of the entity is S_2. The tree of computation spaces is traversed upwards starting from S_2. If during traversal S_1 is encountered, it holds that $S_2 \in \Downarrow S_1$ and the entity is copied.

This solution is inefficient in that the space tree is traversed for each situated entity. A first improvement is that threads need no situatedness-check. This is due to the invariant that all references to threads are downward with respect to the tree.

The remaining situatedness checks can be optimized easily. Before copying of S starts, all spaces in $\uparrow S$ are marked. If during copying a situated entity with home S' is encountered, the check is performed by testing whether S' is marked. A further optimization is to avoid dereferencing, which is required for accessing the home space (this is due to merging, Section 13.4). Dereferencing can be avoided by marking *all* spaces including merged spaces that otherwise are considered transparent during dereferencing.

Example 13.4 (Taking Advantage of Situated Entities). Entities situated in $\uparrow S$ are not copied when cloning S and hence require no memory. This can be utilized by explicitly situating data structures by wrapping them using procedural abstraction. For example,

```
S={NewSpace proc {$ X} Y in {Wait Y} X=Data end}
```

when S is cloned, Data (potentially large) is cloned. This is avoided by

```
fun {GetData} Data end
S={NewSpace proc {$ X} Y in {Wait Y} X={GetData} end}
```

When cloning S, Data is *not* cloned.

Janson describes in [58] how to situate tree constraints. Each node has a reference to the creating space and unification updates the reference accordingly. The technique always incurs memory overhead for all data structures. Its applicability is definitely limited. Typically, data is either used as input to the computations in a space, or data is subject to local computations anyway and the optimization cannot offer any advantage. If the data is used as input, it can be situated if necessary by procedural abstraction.

Implementation Effort. The close relationship to garbage collection keeps the implementation effort for cloning small. The Mozart implementation, for example, uses the same templates for garbage collection and for cloning. The templates are specialized during compile-time for garbage collection and cloning.

13.6 Richer Basic Constraints

This section is concerned with extensions required by richer basic constraints.

13.6.1 Variable Aliasing

For a store with equations between variables, the implementation is extended as follows.

Aliasing Order. Telling a speculative constraint $x_1 = x_2$ where $S_i = \mathcal{H}(x_i)$ and $S_2 < S_1$ requires that the binding is established from x_1 to x_2 ("binding is upward"). If S_1 is the current space, this condition ("bind local to global variable") is essential for entailment, since it ensures the minimality invariant introduced in Section 13.2.3. Smolka and Treinen discuss in [137] that this criteria is sufficient for entailment.

The "binding is upward" condition is needed for stability depending on globally suspended threads. If a thread T suspends on x_1 or x_2, T becomes speculative for both S_1 and S_2. If the link had been established from x_2 to x_1, the scheduler would enter T to the global suspension set of S_1 but not to the global suspension set of S_2. Hence S_2 would be detected as stable, even though T is speculative for S_2.

Supervisor Threads. Script creation for S takes into account that variable pairs $\langle x_1, x_2 \rangle$ can be put into the script. In this case, $\{\mathcal{H}(x_1), \mathcal{H}(x_2)\} \subseteq \uparrow S$ (otherwise, $x_1 = x_2$ is not speculative). Hence, the supervisor thread suspends on both x_1 and x_2.

Script Installation. No threads are woken during script installation. If the script contains a variable pair $\langle x_1, x_2 \rangle$, all threads that suspend on x_1 or x_2 have already been woken when the binding has been established initially.

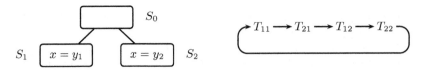

Fig. 13.7. Oscillating threads (Example 13.5).

Example 13.5 (Oscillating Threads). This example demonstrates why it is essential to not wake threads during script installation. Consider a space S_0 with children S_1 and S_2 (Figure 13.7). Each S_i ($i \in \{1,2\}$) contains a speculative binding $x = y_i$ together with threads T_{i1} and T_{i2} that are runnable but when being run will immediately suspend on x. Further suppose that the scheduler executes the threads in order T_{11}, T_{21}, T_{12}, T_{22}.

When T_{11} is executed, S_1 is installed. As said above, T_{11} suspends. Now T_{21} is executed, S_2 is installed, and T_{21} immediately suspends. And now the disaster takes place: T_{12} is run. By installing S_1, the speculative binding is installed which wakes T_{11}. The same happens with execution of T_{22}: by installing S_2, T_{21} is woken again. Now both T_{11} and T_{21} are runnable again. Their execution will in turn make T_{12} and T_{22} runnable ...: S_1 and S_2 never become blocked and thus S is never detected as being stable.

13.6.2 Tree Constraints

Telling a tree constraint can introduce new variable bindings. This is taken into account during script installation. Suppose that S contains the speculative constraint $x = f(y)$ and in a space in $\uparrow S$ the constraint $x = f(n)$ is told. When installing the script for S, the tell $y = n$ is performed. Threads that suspend on y are woken, since $y = n$ is new. If during script installation new variable bindings are possible, threads are woken.

Note that a situation as in Example 13.5 cannot happen. Only pairs $\langle x, v \rangle$ are entered into the script: provided that no new tree constraints are told on x, the next installation of the script does not wake any threads.

13.6.3 Finite Domain Constraints

The domain of a finite domain variable can be repeatedly narrowed. This also holds true for finite set variables and feature constraints. Variables that can be repeatedly constrained are referred to as *constraint-variables.*

Constraint Trailing. So far it was sufficient to record the variable being constrained on the trail. For a constraint-variable also the current constraint (for example, the current domain of a finite domain variable) is stored. Undoing the speculative constraint on a constraint-variable reestablishes the constraint from the trail.

Dually, the constraint on a variable x is stored in the script for a space S. This requires no extension of the script data structure. Assume that $x \in D$. A pair $\langle x, y \rangle$ is put into the script, where y is a new constraint-variable with $\mathcal{H}(y) = S$ and $y \in D$.

Time-Marking. Multiply constraining a constraint-variable x also means to multiply trail x and the constraint associated with x as described above. This is not necessary, as only the initial constraint must be reestablished.

A common technique to avoid multiple entries on the trail for the same variable is *time-stamping*, which has been first considered in CHIP [1, 2]. The idea is as follows: each time a new speculative context is entered (in this context, a space is installed) a global time-stamp is incremented. When a variable is constrained, the variable is marked with the global time-stamp. The variable must only be trailed, if its time-stamp is less than the global time-stamp.

Time-marking is used as a different implementation that requires less memory at a slight expense in runtime for space installation. When a variable is constrained for the first time in the current space, it is marked (just a single bit). If a variable carries a mark, it needs no trailing. When the current space S is deinstalled, the marks of all variables are reset (the marked variables are found in the trail). For the new current space (S's parent) the trail is scanned and all variables on the trail are re-marked. The same technique is used in script installation.

Time-marking is used since speculative constraints on constraint-variables are infrequent. This is different from the motivation for time-stamping, where the trail stores information needed for trailing-based search (Section 14.2 discusses trailing-based search and its comparison to copying).

Variable Aliasing. Variable aliasing must respect situatedness as discussed in Section 13.6.1. This also holds true for constraint-variables. In particular, the case can arise where a constraint-variable must be bound to a non-constrained variable. When considering stability, the technique described by Van Roy, Mehl, and Scheidhauer in [149] to first create a local constraint-variable to which both global variables are bound is incorrect. Würtz considers stability and wrongly proposes this technique for finite domain variables [157].

13.7 Ports

The main issue with message sending to ports across space boundaries is testing sendability. Consider a send operation {Send x y}, where x is determined to a port with home S_1 and S_2 is the current space. If $S_1 = S_2$, no checking is required. Otherwise the constraints on y are checked by constraint graph traversal similar to the graph traversal for cloning. All situated nodes v encountered during traversal are checked whether $\mathcal{H}(v) \leq S_1$, where the same optimized situatedness test as in cloning is used.

13.8 Performance Overview

This section gives an overview of the performance of operations on spaces. An empirical comparison with other constraint programming systems is contained in Section 14.3. Appendix A.2 provides more information on the used software and hardware platform.

Oz Light Operations. Figure 13.8(a) shows the performance of the central operations of Oz Light. These figures serve as comparison to the performance of space operations.

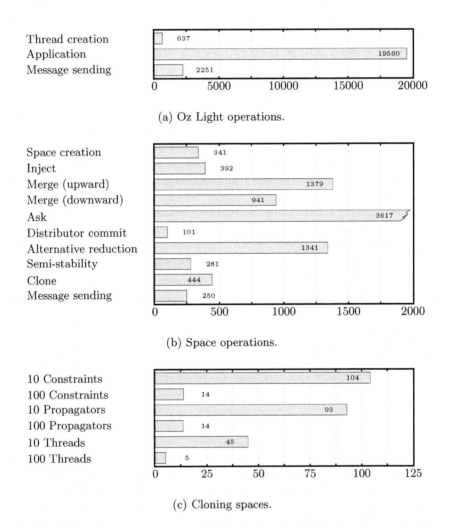

(a) Oz Light operations.

(b) Space operations.

(c) Cloning spaces.

Fig. 13.8. Base performance of operations (in thousand operations per second).

Space Operations. Figure 13.8(b) shows the performance of space operations. The script for "Space creation" and "Inject" contains **skip** as its body. The space used for "Clone" contains its root variable. "Distributor commit" captures creation of a binary distributor and committing to one of its alternatives. "Alternative reduction" reduces a ternary to a binary distributor (Section 4.5.6).

Reducing alternatives of distributors is efficient. This justifies the usage of binarization as discussed in Section 5.3. Message sending for a typical message (a tuple containing a port, an atom, and a list with two integers) is one order of magnitude slower with testing sendability.

Cloning Spaces. The time needed to clone a space depends on the number of variables, constraints, threads, and propagators situated in the space. Figure 13.8(c) shows the number of spaces that can be cloned per second depending on the space's content. The spaces used are as follows: "n Constraints" contain n domain constraints for the domain $\{1, \ldots, 100\}$, "n Propagators" contain n binary propagators for $x \neq y$, and "n Threads" contain n threads that synchronize on a single variable.

The numbers show that cloning is linear in the number of basic constraints, propagators, and threads. Cloning gets more efficient as the spaces contain more content. This is due to the fact that the overhead for cloning the data structures representing the space itself remains constant regardless of its content.

Search Engines. The numbers in Figure 13.9 give the performance of basic search engines in thousand nodes explored per second. All engines explore a complete binary search tree with $2^{16} - 1$ nodes. For "BAB (failed)" all leaves of the search tree are failed, whereas for "BAB (solutions)" all leaves are solutions. While the former gives an upper bound on the performance, the latter is a lower bound.

"Explorer (Hidden)" gives the performance of the Explorer (Chapter 8) for exploring the search tree, "Explorer (Full)" includes drawing of the entire search tree, without hiding any part of it.

The numbers yield important information on the minimal size of a search problem that can be tackled efficiently using space-based search engines. Due to the little overhead of search engine, they are efficient enough to be employed for even small problems.

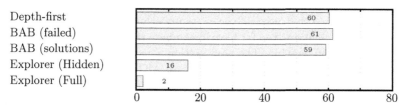

Fig. 13.9. Base performance of search engines (in thousand nodes per second).

14. Other Approaches to Search

This chapter compares space-based search with approaches to search found in other constraint programming systems. A general discussion is followed by a detailed comparison of copying with trailing as the dominating implementation technique for search. Empirical evaluation demonstrates that copying-based search together with recomputation is competitive with trailing-based search and is superior for large examples.

14.1 Other Constraint Programming Systems

Most of todays constraint programming systems are constraint logic programming systems (CLP) that evolved from Prolog and inherited Prolog's search capabilities: CHIP [2, 31], Eclipse [151], clp(FD) [23] and its successor GNU Prolog [29], and SICStus [15], just to name a few. Also cc(FD) [145] shares the approach to search taken by CLP-based systems. Jaffar and Maher provide an overview on CLP in [57].

Screamer [131] is based on Common Lisp and supports finite domain constraints and backtracking search similar to Prolog. Claire [18] is a programming language for set-based and rule-based programming. Search is supported by a versioning mechanism that allows backtracking search. SALSA [72] is a language for the specification of search algorithms that cover distribution strategies for tree search as well as neighborhood-based search (local search). SALSA requires a host language that supports search (for example, Claire) as compilation target. ALMA-O [4] extends Modula-2 by choice points which are resolved by backtracking-based search.

ILOG Solver [55, 110, 111] is a constraint programming library that uses C++ as its host language. Solver provides finite domain constraints, finite set constraints, and constraints over real numbers. OPL [142, 144] is a constraint modeling language that uses Solver as its underlying execution platform.

All systems mentioned so far have in common that they are based on trailing rather than on copying. The next section compares copying and trailing.

Close relatives to computation spaces are AKL and Curry. AKL [44, 58] has pioneered encapsulated search and stability. An extension of AKL with finite domain constraints is described in [12]. Curry [40] is a functional logic language that also provides encapsulated search [41]. Encapsulated search in

C. Schulte: Programming Constraint Services, LNAI 2302, pp. 143–152, 2002.
© Springer-Verlag Berlin Heidelberg 2002

Curry has adopted a variant of the solve combinator [128, 129] (Section 4.7 discusses the solve combinator and its limitations). The combinator in Curry offers distributors with a dynamic number of alternatives which has not been possible with the originally proposed solve combinator.

Predefined Search Strategies. All CLP-based languages support single- and all-solution search. Best-solution search is controlled by a single cost variable and amounts to search for a solution with smallest or largest cost. CLP-based systems offer an interactive toplevel that allows the user to prompt for multiple solutions. The toplevel cannot be used within programs. Eclipse provides visual search through the Grace tool [79] (Section 8.6).

Solver (and hence OPL) additionally offers LDS [48], DDS [153], and IDFS [80] (Section 5.6). Best-solution search in Solver also uses a cost variable. To avoid recomputation of the best solution, the program must be modified to explicitly store solutions. Search in Solver is incremental in that solutions can be computed on request.

Best-solution search based on a cost-variable requires to map the ordering between solutions as possible with spaces to a single value. This can result in complicated solutions that might compromise propagation (think of mapping two variables to a single cost variable to express a lexicographic order).

Programming Exploration. Only Solver (and OPL) and Curry offer support for programming exploration, where Curry offers the same programming model as the solve combinator. Programming exploration in Solver is based on limits and node evaluators [100, 144]. Programmable limits allow to stop exploration (time limit, for example). Node evaluators map search tree nodes to priorities. Node priorities determine the exploration order of nodes. Additionally, a special priority discards nodes.

Solver supports switching between arbitrary nodes in the search tree by full recomputation. For example, best-first search needs to switch between arbitrary nodes. To limit the amount of switching, Solver uses an additional threshold value. Only if the cost improvement exceeds the threshold, nodes are switched. This results in an approximation of best-first search. Fully interactive exploration is not feasible with full recomputation.

Encapsulation and Control. AKL shares encapsulation and stability with computation spaces. Curry offers encapsulation and a simpler and a more limited control regime than stability: execution stops as soon as a speculative constraint is told. Apart from the limitations that are caused by the solve combinator (Section 4.7), this restriction resembles the independence restriction of Chapter 4 and excludes full compositionality and speculative execution, as for example needed for programming combinators (Chapter 11).

Solver controls and encapsulates search by a manager. Multiple independent managers are possible but cannot be nested with automatic propagation of constraints. Managers support two different modes of operation: edit and search mode. Propagation and search is disabled during edit mode which allows the setup of constraint problems, including removal of constraints.

14.2 Comparison with Trailing

Search demands that previous computation states must possibly be available at a later stage of computation. A system must take precaution by either memorizing states or by means to reconstruct them. States are memorized by *copying*. Techniques for reconstruction are *trailing* and *recomputation*. While recomputation computes everything from scratch, trailing records for each state-changing operation the information necessary to undo its effect.

Copying offers advantages with respect to expressiveness: multiple nodes of a search tree are available simultaneously for further exploration. This is essential for concurrent, parallel, breadth-first, and user-defined search strategies. Implementation can be simpler, since copying is independent of operations and is only concerned with data structures.

On the other hand, copying needs more memory and might be slower since full copies of the computation states are created. Hence, it is not at all clear whether copying is competitive to trailing or not.

This section shows that copying is indeed competitive and that it offers a viable alternative to trailing for the implementation of constraint programming systems. It is clarified how much more memory copying needs. It is examined for which problems copying is competitive with respect to runtime and memory. For large problems with deep search trees this section confirms that copying needs too much memory. It is shown that in these cases recomputation can decrease memory consumption considerably, even to a fraction of what is needed by trailing.

14.2.1 Expressiveness

The main difference in expressiveness between copying and trailing is the number of nodes simultaneously available for exploration. With copying, all nodes that are created as copies are directly ready for further exploration. With trailing, exploration can only continue at a single node at a time. In principle, trailing does not exclude exploration of multiple nodes. However, they can be explored in an interleaved fashion only and switching between nodes is a costly operation.

Having more than a single node available for exploration is essential to search strategies like concurrent, parallel (Chapter 9), or best-first (Section 5.7). The same property is crucial for user-defined interactive exploration as implemented by the Oz Explorer (Chapter 8).

Resource Model. Copying essentially differs from trailing with respect to space requirements in that it is *pessimistic*: while trailing records changes exactly, copying makes the safe but pessimistic assumption that everything will change. On the other hand, trailing needs to record information on what changes as well as the original state of what is changed. In the worst case — the entire state is changed — this might require more memory than copying. This discussion makes clear that a meaningful comparison of the space requirements for trailing and copying is only possible by empirical investigations, which are carried out in Section 14.2.5.

14.2.2 Implementation Issues

This section gives a short discussion of the main implementation concepts and their properties in copying- and trailing-based systems. The most fundamental distinction is that *trailing*-based systems are concerned with *operations* on data structures while *copying*-based systems are concerned with the *data structures* themselves.

Copying. Copying needs for each data structure a routine that creates a copy and recursively copies contained data structures. A system that features a copying garbage collector already provides almost everything needed to implement copying (see Section 13.5.2).

By this, all operations on data structures are independent of search with respect to both design and implementation. This makes search in a system an orthogonal issue. Development of the Mozart system has proven this point: it was first conceived and implemented without search and only later search has been added.

Trailing. A trailing-based system uses a trail to store undo information. Prior to performing a state-changing operation, information to reconstruct the state is stored on the trail. In a concrete implementation, the state changing operations considered are updates of memory locations. If a memory update is performed, the location's address and its old content is stored on the trail. This kind of trail is referred to as *single-value trail*. Starting exploration from a node puts a mark on the trail. Undoing the trail restores all memory locations up to the previous mark. This is essentially the technology used in Warren's Abstract Machine [5, 154].

In the context of trailing-based constraint programming systems two further techniques come into play:

Time-Stamping. With finite domains, for example, the domain of a variable can be narrowed multiply. However it is sufficient to trail only the original value, because intermediate values need no restauration: each location needs to appear at most once on the trail. Otherwise memory consumption is no longer bounded by the number of changed locations but by the number of state-changing operations performed. To ensure this property, time-stamping is used: as soon as an entity is trailed, the entity is stamped to prevent it from further trailing until the stamp changes again. Note that time-stamping concerns both the operations and the data structures that must contain the time-stamp. Section 13.6.3 discusses time-marking as an alternative to time-stamping.

Multiple-Value Trail. A single-value trail needs $2n$ entries for n changed locations. A multiple-value trail uses the optimization that if the contents of $n > 1$ successive locations are changed, $n + 2$ entries are added to the trail: the location's address, n itself, and n entries for the locations' values.

For a discussion of time-stamps and a multiple-value trail in the context of the CHIP system, see [1, 2]. A general but brief discussion of issues related to the implementation of trailing-based constraint programming systems can be found in [57].

Trailing requires that all operations are search-aware: search is not an orthogonal issue to the rest of the system. Complexity in design and implementation is increased: it is a matter of fact that a larger part of a system is concerned with operations rather than with basic data structure management. A good design that encapsulates update operations will avoid most of the complexity. To take advantage of multiple-value trail entries, however, operations require special effort in design and implementation.

Trailing for elaborated data structures can become quite complex. Consider as an example adding an element to a dictionary with subsequent reorganization of the dictionary's hash table. Here the simple model that is based on trailing locations might be unsuited, since reorganizing data structures alters a large number of locations. In general, copying offers more freedom of rearranging data structures. Müller and Würtz discuss this issue in the context of finite domain constraints in [86].

The discussion in this section can be summarized as follows. A system that features a copying garbage collector already supports the essential functionality for copying. For a system that does not require a garbage collector, implementing trailing might be as easy or possibly easier depending on the number and complexity of the operations.

14.2.3 Criteria and Examples

This section introduces constraint problems that serve as examples for the empirical analysis and comparison. The problems are well known and are chosen to be easily portable to several constraint programming systems (Section 14.3).

The main characteristics of the problems are listed in Appendix A.1. Besides of portability and simplicity they cover a broad range with respect to the following criteria.

Problem Size. The problems differ in size, that is in the number of variables and constraints, and in the size of constraints (that is the number of variables each constraint is attached to). With copying, the size of the problem is an important parameter: it determines the time needed for copying. Additionally, it partly determines the memory requirements (which is also influenced by the search tree depth). Hence, large problem sizes can be problematic with copying.

Propagation Amount. Strong propagation narrows a large number of variables. This presupposes a large number of propagation steps, which usually coincides with state changes of a large number of constraints. The amount of propagation determines how much time and memory trailing requires: the stronger the propagation, the more of the state is changed. The more of the state changes, the better it fits the pessimistic assumption "everything changes" that underlies copying.

Search Tree Depth. The depth of the search tree determines partly the memory requirements for both trailing and copying. Deep search trees are a bad case for trailing and even more for copying due to its higher memory requirements.

Exploration Completeness. How much of the search tree is explored. A high exploration completeness means that utilization of the precaution effort undertaken by copying or trailing is high.

The criteria are mutually interdependent. Of course, the amount of propagation determines the depth of the search tree. Also search tree depth and exploration completeness are interdependent: If the search tree is deep, exploration completeness is definitely low. Due to the exponential number of nodes, only a small part of the tree can be explored.

Familiar benchmark programs are preferred over more realistic problems such as scheduling or resource allocation. The reason is that the programs are also intended for comparing several constraint programming systems. Choosing simple constraints ensures that the amount of constraint propagation is the same with all compared systems.

Evaluations of Oz that specifically address scheduling problems are [156]. Reports on successful applications using copying-based search are [50, 52, 54, 120].

14.2.4 Copying

This section presents and analyses runtime and memory requirements for Mozart. Appendix A.2 contains more information on hardware and software platforms.

Table 14.1 displays the performance of the example programs. The fields "Copy" and "GC" give the percentage of runtime that is spent on copying and garbage collection, the field "CGC" displays the sum of both fields. The field "Max" contains the maximal amount of memory used in Kilobytes, that is how much memory must at least be available in order to solve the problem.

The numbers clarify that for all but the large problems 100-Queens and 18-Knights the amount of time spent on copying and garbage collection is around one fourth of the total runtime. In addition, the memory requirements are moderate. This demonstrates that for problems with small and medium size copying does neither cause memory nor runtime problems. It can be expected that for these problems copying is competitive.

Table 14.1. Runtime and memory performance of example programs.

Example	Time	Copy	GC	CGC	Max
	msec	%	%	%	KB
Alpha	1975	20.2	0.0	20.2	19
10-Queens	739	33.5	0.0	33.5	20
10-S-Queens	572	21.4	0.0	21.4	7
100-Queens	868	49.3	18.7	68.0	21873
100-S-Queens	26	28.6	0.0	28.6	592
Magic	606	13.3	14.1	27.4	6091
18-Knights	5659	44.2	22.3	66.5	121557

On the other hand, the numbers confirm that *copying alone for large problems with deep search trees is unsuited*: up to two third of the runtime is spent on memory management and memory requirements are prohibitive. The considerable time spent on garbage collection is a consequence of copying: the time used by a copying garbage collector is determined by the amount of used memory.

The two different implementations of *n*-Queens exemplify that copying gets considerably better for problems where a large number of small propagators is replaced by a small number of equivalent global propagators.

14.2.5 Copying versus Trailing

As discussed before, one of the most essential questions in comparing trailing and copying is: how pessimistic is the assumption "everything changes" that underlies copying. An answer seems to presuppose two systems that are identical with the exception of trailing or copying. Implementing two competitive systems is not feasible.

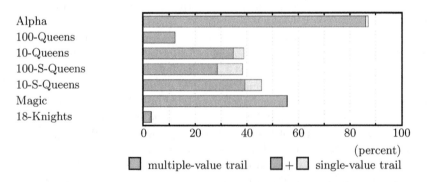

Fig. 14.1. Memory use of trailing versus copying.

Instead, the memory requirements of a trailing implementation are computed from the requirements of a copying implementation as follows. Before constraint propagation in a node N begins, a bitwise copy of the memory area occupied by N is created. After constraint propagation has finished, this area is compared to the now changed area occupied by N. The altered locations are those that a trailing system must have trailed.

Figure 14.1 shows the percentage of memory needed by a trailing implementation compared to a copying implementation. The total length of bars depicts the percentage needed by a single-value trail, whereas the dark-colored bar represents the need of a multiple-value trail implementation.

The percentage figures for the multiple-value trail are lower bounds again. Locations that are updated by separate single update operations might happen to be successive even though an implementation cannot take advantage of this fact. It is interesting to note that a multiple-value trail offers some improvement only for 10-S-Queens and 100-S-Queens (around 10%). Otherwise, its impact is quite limited (less than 2%).

The observation that for large problems with weak propagation (100-Queens and 18-Knights) trailing improves by almost up to two orders of magnitude coincides with the observation made with respect to the memory requirements in Section 14.2.4. For the other problems the memory requirements are in the same order of magnitude and trailing roughly halves them.

What is not captured at all by the comparison's method is that other design decisions for propagators would have been made to take advantage of trailing, as has already been argued in Section 14.2.2.

14.2.6 Recomputation versus Trailing

Fixed recomputation (Section 7.3) uses less memory than trailing. Figure 14.2 shows the percentage of memory that fixed recomputation takes in comparison to the memory needed by trailing.

Trailing and copying are pessimistic in that they make the assumption that each node needs reconstruction. Recomputation, in contrast, makes the optimistic assumption that no node requires later reconstruction. For search trees that contain few failed nodes, the optimistic assumption fits well. In particular, problems with very deep search trees can profit from the optimistic assumption, since exploration completeness is definitely low.

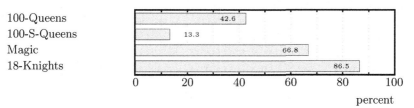

Fig. 14.2. Memory use of fixed recomputation versus trailing.

14.3 System Comparison

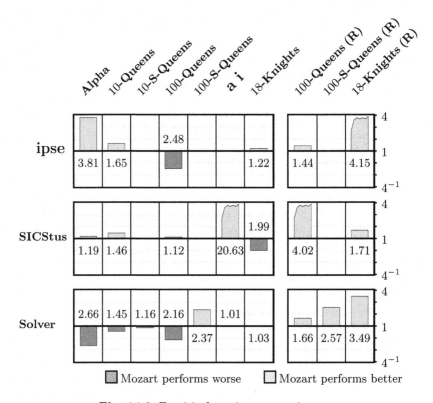

Fig. 14.3. Empirical runtime comparison.

This section compares Mozart, a copying-based system, with several trailing-based systems. Appendix A.2 contains more information on the used software and hardware platforms. The point to compare systems is to demonstrate that a copying-based system can be competitive with trailing-based systems.

All systems support Alpha, 10-Queens, 100-Queens, and 18-Knights. The propagators that are used for 10-S-Queens and 100-S-Queens are available in Mozart and Solver only. Eclipse does not support the exactly-constraint that is used in Magic.

Figure 14.3 shows a relative performance comparison of Mozart with Eclipse, SICStus, and Solver. The figures to the left are without recomputation, the figures to the right use adaptive recomputation. As initial value for the MRD 10% of the search tree depth is used (Section 7.5 discusses the little impact of the initial MRD). A number of f below the middle line together with a light gray box means that Mozart performs f-times better. Otherwise, the other system performs f-times better than Mozart.

The figures clearly indicate that a system based on copying is competitive as it comes to runtime. Even for problems that profit from recomputation, performance is still competitive without recomputation. In general, this is of course only true if the available memory is sufficient.

The numbers for Mozart with adaptive recomputation show that copying together with recomputation for large problems and deep search trees outperform trailing-based systems. An important point is that adaptive recomputation is *automatic* and does not require any parameter tuning.

Impact of Finite Domain Implementation. The runtimes of course do not depend only on the systems' search capabilities, but also on their finite domain implementation. It has been tried to keep the examples' implementations for the different systems as similar as possible. Even if a system provides special constraints for a particular example, the programs do not take advantage:

- 10-Queens and 10-S-Queens can be implemented more efficiently in SICStus by directly using indexicals as provided by the underlying constraint solver [14].
- Both Eclipse and SICStus implement domains as list-of-intervals rather than as bit-vectors and list-of-intervals as Mozart does: this explains why Mozart is quite efficient for 10-Queens and 10-S-Queens in comparison.
- The performance of Magic for SICStus is due to a naive implementation of the exactly-constraint [13].

15. Conclusion

This chapter summarizes the main contributions of this book and presents concrete ideas for future work.

15.1 Main Contributions

This book develops computation spaces as simple programming abstractions for constraint services at a high-level. It presents a tight integration of spaces into a concurrent programming language. The tight integration is proven to ease programming and integration into todays concurrent and distributed computing environments. The appropriateness of spaces is demonstrated by application to state-of-the-art search engines, to entirely new search engines, and to composable constraint combinators. A simple yet efficient implementation is presented that competes with today's best commercially available constraint programming systems.

Search. Spaces cover state-of-the-art search engines, such as plain, best-solution, and best-first search. They cover new and highly relevant search engines such as visual and interactive search and parallel search utilizing the computational power of networked computers. Spaces allow for succinct programs which are amenable to generalization. Examples are the generalization of branch-and-bound to prune search and search engines with explicit state to concurrent search engines. The Explorer and parallel search engines exemplify the rich support for controlling search.

Recomputation. The combination of recomputation and copying provides search engines that offer a fundamental improvement over trailing-based search for truly large problems. This book shows that adaptive recomputation is an excellent technique for solving large problems. It establishes the competitiveness of copying by a rigid comparison with trailing.

Encapsulation and Integration. Computation spaces provide encapsulation to speculative constraint-based computations, a must for the integration of constraint programming into todays concurrent and distributed computing infrastructure. Encapsulation is achieved by a tight integration of spaces into the concurrent programming language Oz together with stability as powerful control regime. The tight integration is shown to be advantageous. It is

C. Schulte: Programming Constraint Services, LNAI 2302, pp. 153–156, 2002.
© Springer-Verlag Berlin Heidelberg 2002

the tight integration into a programming language that accounts for ease of programming. It is the tight integration with concurrency that enables programming of composable constraint combinators and parallel search engines.

Coordinating Speculative Computations. Ports as well-established communication mechanism are generalized to allow global coordination by communication with speculative computations while obeying encapsulation. Active services based on ports provide a familiar programming model resembling remote procedure call (RPC) and remote method invocation (RMI).

Composable Combinators. Composable combinators (also known as deep-guard combinators) are shown to have a surprisingly simple implementation with spaces. They are show-cases for concurrency and encapsulation. The fresh look at combinators by simple composition from abstractions contributed new insights such as how to employ stability to detect programming errors due to stuck computations.

Implementation. The implementation is factored into orthogonal support for constraint stores, stability, space operations, and search. Scripting is used as a technique that requires few and conservative extensions. Supervisor threads effectively decouple constraint-dependent aspects from the rest of the implementation. Copying leads to a simple implementation of search that takes little effort.

Production Quality System. Spaces and services programmed from spaces have already proven their usefulness and maturity to many users of the Mozart implementation of Oz. The Mozart implementation is a production quality system that is successfully used in large applications. It offers unique tools like the Oz Explorer for the development and distributed search engines for the deployment of applications due to the material developed in this book.

Impact. Some ideas in this book have already proven their impact. The CHIP search tree tool [130] has been inspired by the Explorer. Encapsulated search in Curry is based on a variant of the solve combinator [41].

On a more general level, I am convinced that future constraint programming systems will support the programming of search engines. I am also convinced that the combination of copying and recomputation will establish itself as a serious alternative for implementing search that at least matches the virtues of trailing.

15.2 Future Work

Formal Model. The informal model for computation spaces presented in Chapters 4 and 10 serves as starting point on how to use and implement spaces. This book provides evidence that spaces are indeed useful and can be implemented efficiently. Given this, the investment into a formal model for spaces seems justified. The hope is that such a model exists and that it is useful to formalize and prove interesting properties on spaces. A particularly interesting and challenging question is whether the implementation with its central invariants can be proven correct.

Libraries Instead of Languages. This monograph introduces computation spaces as abstractions that support the programming of constraint services. Spaces are tightly integrated into a concurrent programming language. This integration is undoubtedly useful as witnessed by application of spaces to composable constraint combinators and parallel search engines.

However, a valid and interesting question is: what if the language does not provide concurrency and implicit synchronization? How can the programming capabilities provided by spaces be transferred to a library in a programming language neutral way? The library approach to constraint programming has been proven successful by ILOG Solver [110]. Further attempts in this direction are Figaro [53] and CHOCO [71].

Dependency Recording. The search strategies considered in this book do not record and utilize information why a particular node in the search tree failed. So-called *lookback schemes* [27, 65] analyze information found in the search tree's nodes and continue exploration at a node such that the same conflict is not encountered again. This form of exploration requires elaborate control and recording of dependency information.

Spaces provide elaborate control. It is interesting to understand what additional space primitives are required for lookback schemes. It is in particular interesting how dependency recording, which depends on the constraint domain, can be integrated with spaces while keeping their domain-independence.

Resource Adaptive Recomputation. The book has demonstrated the great potential of adaptive recomputation for solving truly large problems. Adaptive recomputation exclusively bases its decision whether to copy or recompute on the shape of the search tree. The costs associated with recomputation and copying for a particular problem are not considered. The hope is that by taking these costs into account, search efficiency can be further improved.

Search Factories. The presentation of search engines in this book individually covers various strategies and programming techniques: parallelism; recomputation strategies; single, all, and best solution search; visualization; interactivity. From a users perspective it is desirable that these features can be orthogonally combined by a search factory: the factory returns a custom-made search engine that has the features required by a particular application. While increasing usability, this idea can help to understand what are the minimal abstractions needed to support a particular feature of a search engine.

Chew, Henz, and Ng describe a search toolkit in [20] that allows the orthogonal combination of some of the features introduced in this book. However, the more challenging features such as parallelism and different recomputation strategies are not covered.

Reusing Parallel Engines. One of the main goals for parallel search engines in Chapter 9 has been a reusable design. That the design is indeed reusable and delivers good performance on shared memory multiprocessors has not yet been assessed. Popov is currently working towards an implementation of Oz that provides thread-based parallelism on multi-processor machines [107]. This implementation will allow to check whether this claim holds true. Ideally, the concurrent search engine should run without any modification and deliver good speedup.

Generational Garbage Collection. A copying garbage collector is a particularly bad choice for copying-based search. Spaces that are copied by cloning will be copied several times by the garbage collector. This accounts for excessive runtime spent on memory management. This is worsened by the fact that copies created for nodes near to the root of the search tree tend to live for a long time. Moreover, these copies do not change.

These facts make a classical case for generational garbage collection [155, 61]. With generational garbage collection, the memory areas that change more often are collected more often. Areas that contain data that changes infrequently are collected infrequently. The hope is that by generational garbage collection the time spent on garbage collection can be dramatically decreased.

A. Benchmark Problems and Platforms

This appendix contains information about examples and platforms used for evaluation.

A.1 Benchmark Problems

This section describes the example constraint problems. Their main characteristics are listed in Table A.1. All are familiar benchmark problems.

Alpha. Alpha is the well-known cryptoarithmetic puzzle: assign variables a, c, ..., z distinct numbers between 1 and 26 such that 25 equations hold.

100-Queens, 100-S-Queens, 10-Queens, and 10-S-Queens. For the n-Queens puzzle (place n queens on a $n \times n$ chess board such that no two queens attack each other) two different implementations are used.
The naive implementation (n-Queens) uses $O(n^2)$ disequality constraints. This is contrasted by a smarter program (n-S-Queens) that uses three propagators for the same constraints.

Magic. The Magic puzzle is to find a magic sequence s of 500 natural numbers, such that $0 \leq x_i \leq 500$ and i occurs in s exactly x_i times. For each element of the sequence an exactly-constraint (ranging over all x_i) on all elements of the sequence is used. The elements are enumerated in increasing order following a splitting strategy.

18-Knights. The goal in 18-Knights is to find a sequence of knight moves on a 18×18 chess board such that each field is visited exactly once and that the moves return the knight to the starting field. The knight starts at the lower left field.

Photo. This example is presented in Chapter 8, although a larger set of persons and preferences is used (9 persons and 17 preferences).

Bridge. Bridge is a small and well-known scheduling example [30]. It requires additional constraints apart from the usual precedence and resource constraints.

MT10, MT10A, and MT10B. These are variants of the 10×10 job-shop scheduling problem due to Muth and Thompson [94]. All variants use `Schedule.serialized` (edge-finding) as serialization-propagator. MT10 uses `Schedule.firstsLastsDist`, MT10A uses `Schedule.lastsDist`,

Table A.1. Characteristics of example programs.

Example	Distr.	Fail.	Sol.	Depth	Var.	Constr.
100-Queens	115	22	1	97	100	14850
100-S-Queens	115	22	1	97	100	3
Magic	13	4	1	12	500	501
18-Knights	266	12	1	265	7500	11205

(a) Single-solution search.

Example	Distr.	Fail.	Sol.	Depth	Var.	Constr.
Alpha	7435	7435	1	50	26	21
10-Queens	6665	5942	724	29	10	135
10-S-Queens	6665	5942	724	29	10	3

(b) All-solution search.

Example	Distr.	Fail.	Sol.	Depth	Var.	Constr.
Bridge	150	148	3	28	198	313
Photo	5471	5467	5	27	61	54
MT10	16779	16701	79	91	102	121
MT10A	17291	17232	60	91	102	121
MT10B	137011	136951	61	91	102	121

(c) Best-solution search (BAB).

and MT10B uses `Schedule.firstsDist` as resource-oriented serializer More information on the serialization propagator and the resource-oriented serializer can be found in [33, Chapter 6].

A.2 Sequential Platform

All numbers but those for distributed search engines in Chapter 9 have been made on a standard personal computer with a 700 MHz AMD Athlon and 256 Megabytes of main memory using RedHat Linux 7.0 as operating system. All times have been taken as wall time (that is, absolute clock time), where the machine was unloaded: difference between wall and actual process time is less than 5%.

The following systems were used: Mozart 1.2.0, Eclipse 5.1.0, SICStus Prolog 3.8.5, and ILOG Solver 5.000.

The numbers presented are the arithmetic mean of 25 runs, where the coefficient of deviation is less than 5% for all benchmarks and systems.

A.3 Distributed Platform

The performance figures presented in Chapter 9 used a collection of standard personal computers running RedHat Linux 6.2 connected by a 100 MB Ethernet. The combination of computers for varying number of workers is shown in Table A.2. The manager has always been run on computer a.

Table A.2. Computers used for evaluation distributed search engines.

Computer	Processors	Memory
a	2×400 MHz Pentium II	256 MB
b	2×400 MHz Pentium II	512 MB
c – f	1×466 MHz Celeron	128 MB

(a) Hardware.

Workers	Combination
1	a
2	a,b
3	a,b,c
4	a,b,c,d
5	a,b,c,d,e
6	a,b,c,d,e,f

(b) Combinations.

References

1. Abderrahamane Aggoun and Nicolas Beldiceanu. Time Stamps Techniques for the Trailed Data in Constraint Logic Programming Systems. In S. Bourgault and M. Dincbas, editors, *Actes du Séminaire 1990 de programmation en Logique*, pages 487–509, Trégastel, France, May 1990. CNET, Lannion, France.
2. Abderrahamane Aggoun and Nicolas Beldiceanu. Overview of the CHIP compiler system. In Benhamou and Colmerauer [10], pages 421–437.
3. Abderrahamane Aggoun, David Chan, Pierre Dufresne, Eamon Falvey, Hugh Grant, Warwick Harvey, Alexander Herold, Geoffrey Macartney, Micha Meier, David Miller, Shyam Mudambi, Stefano Novello, Bruno Perez, Emmanuel Van Rossum, Joachim Schimpf, Kish Shen, Periklis Andreas Tsahageas, and Dominique Henry de Villeneuve. ECLiPSe 5.0. User manual, IC Parc, London, UK, November 2000.
4. Krzysztof R. Apt, Jacob Brunekreef, Vincent Partington, and Andrea Schaerf. Alma-O: An imperative language that supports declarative programming. *ACM Transactions on Programming Languages and Systems*, 20(5):1014–1066, September 1998.
5. Hassan Aït-Kaci. *Warren's Abstract Machine: A Tutorial Reconstruction*. Logic Programming Series. The MIT Press, Cambridge, MA, USA, 1991.
6. Hassan Aït-Kaci and Roger Nasr. Integrating logic and functional programming. *Journal of Lisp and Symbolic Computation*, 2(1):51–89, 1989.
7. Rolf Backofen. Constraint techniques for solving the protein structure prediction problem. In Maher and Puget [73], pages 72–86.
8. Rolf Backofen and Sebastian Will. Excluding symmetries in constraint based search. In Jaffar [56], pages 73–87.
9. Nicolas Beldiceanu, Warwick Harvey, Martin Henz, François Laburthe, Eric Monfroy, Tobias Müller, Laurent Perron, and Christian Schulte. Proceedings of TRICS: Techniques foR Implementing Constraint programming Systems, a post-conference workshop of CP 2000. Technical Report TRA9/00, School of Computing, National University of Singapore, 55 Science Drive 2, Singapore 117599, September 2000.
10. Frédéric Benhamou and Alain Colmerauer, editors. *Constraint Logic Programming: Selected Research*. The MIT Press, Cambridge, MA, USA, 1993.
11. Maurice Bruynooghe, editor. *Logic Programming: Proceedings of the 1994 International Symposium*. The MIT Press, Ithaca, NY, USA, November 1994.
12. Björn Carlson. *Compiling and Executing Finite Domain Constraints*. PhD thesis, SICS Swedish Institute of Computer Science, SICS Box 1263, S-164 28 Kista, Sweden, 1995. SICS Dissertation Series 18.
13. Mats Carlsson. Personal communication, May 2000.

14. Mats Carlsson, Greger Ottosson, and Björn Carlson. An open-ended finite domain constraint solver. In Hugh Glaser, Pieter H. Hartel, and Herbert Kuchen, editors, *Programming Languages: Implementations, Logics, and Programs, 9th International Symposium, PLILP'97*, volume 1292 of *Lecture Notes in Computer Science*, pages 191–206, Southampton, UK, September 1997. Springer-Verlag.

15. Mats Carlsson, Johan Widén, Johan Andersson, Stefan Andersson, Kent Boortz, Hans Nilsson, and Thomas Sjöland. SICStus Prolog user's manual. Technical report, Swedish Institute of Computer Science, Box 1263, 164 28 Kista, Sweden, 1993.

16. Manuel Carro, Luis Gómez, and Manuel Hermenegildo. Some paradigms for visualizing parallel execution of logic programs. In David S. Warren, editor, *Proceedings of the Tenth International Conference on Logic Programming*, pages 184–200, Budapest, Hungary, June 1993. The MIT Press.

17. Yves Caseau. Using constraint propagation for complex scheduling problems: Managing size, complex resources and travel. In Smolka [135], pages 163–166.

18. Yves Caseau, François-Xavier Josset, and François Laburthe. CLAIRE: Combining sets, search and rules to better express algorithms. In De Schreye [26], pages 245–259.

19. Jacques Chassin de Kergommeaux and Philippe Codognet. Parallel logic programming systems. *ACM Computing Surveys*, 26(3):295–336, September 1994.

20. Tee Yong Chew, Martin Henz, and Ka Boon Ng. A toolkit for constraint-based inference engines. In Pontelli and Costa [105], pages 185–199.

21. W. F. Clocksin. Principles of the DelPhi parallel inference machine. *The Computer Journal*, 30(5):386–392, 1987.

22. W. F. Clocksin and H. Alshawi. A method for efficiently executing horn clause programs using multiple processors. *New Generation Computing*, 5:361–376, 1988.

23. Philippe Codognet and Daniel Diaz. Compiling constraints in clp(FD). *The Journal of Logic Programming*, 27(3):185–226, June 1996.

24. Alain Colmerauer. Equations and inequations on finite and infinite trees. In *Proceedings of the 2nd International Conference on Fifth Generation Computer Systems*, pages 85–99, 1984.

25. James M. Crawford. An approach to resource constrained project scheduling. In George F. Luger, editor, *Proceedings of the 1995 Artificial Intelligence and Manufacturing Research Planning Workshop*, Albuquerque, NM, USA, 1996. The AAAI Press.

26. Danny De Schreye, editor. *Proceedings of the 1999 International Conference on Logic Programming*. The MIT Press, Las Cruces, NM, USA, November 1999.

27. Rina Dechter. Enhancement schemes for constraint processing: Backjumping, learning and cutset decomposition. *Artificial Intelligence*, 41(3):273–312, January 1990.

28. Pierre Deransart, Manuel V. Hermenegildo, and Jan Małuszyński, editors. *Analysis and Visualization Tools for Constraint Programming: Constraint Debugging*, volume 1870 of *Lecture Notes in Computer Science*. Springer-Verlag, Berlin, Germany, 2000.

29. Daniel Diaz and Philippe Codognet. GNU prolog: Beyond compiling Prolog to C. In Pontelli and Costa [105], pages 81–92.

30. Mehmet Dincbas, Helmut Simonis, and Pascal Van Hentenryck. Solving Large Combinatorial Problems in Logic Programming. *The Journal of Logic Programming*, 8(1-2):74–94, January-March 1990.

31. Mehmet Dincbas, Pascal Van Hentenryck, Helmut Simonis, Abderrahamane Aggoun, Thomas Graf, and F. Berthier. The constraint logic programming language CHIP. In *Proceedings of the International Conference on Fifth Generation Computer Systems FGCS-88*, pages 693–702, Tokyo, Japan, December 1988.

32. Denys Duchier and Claire Gardent. A constraint-based treatment of descriptions. In H. C. Bunt and E. G. C. Thijsse, editors, *Third International Workshop on Computational Semantics (IWCS-3)*, pages 71–85, Tilburg, NL, January 1999.

33. Denys Duchier, Leif Kornstaedt, Tobias Müller, Christian Schulte, and Peter Van Roy. *System Modules*. The Mozart Consortium, www.mozart-oz.org, 1999.

34. Denys Duchier, Leif Kornstaedt, and Christian Schulte. *Application Programming*. The Mozart Consortium, www.mozart-oz.org, 1999.

35. Denys Duchier, Leif Kornstaedt, Christian Schulte, and Gert Smolka. A higher-order module discipline with separate compilation, dynamic linking, and pickling. Technical report, Programming Systems Lab, Universität des Saarlandes, www.ps.uni-sb.de/papers, 1998. Draft.

36. Michael Fröhlich and Mattias Werner. Demonstration of the interactive graph visualization system daVinci. In Tamassia and Tollis [139], pages 266–269.

37. Thom Frühwirth. Constraint handling rules. In Podelski [101], pages 90–107.

38. Carmen Gervet. Interval propagation to reason about sets: Definition and implementation of a practical language. *Constraints*, 1(3):191–244, 1997.

39. Gopal Gupta and Bharat Jayaraman. Analysis of OR-parallel execution models. *ACM Transactions on Programming Languages and Systems*, 15(4):659–680, October 1993.

40. Michael Hanus. A unified computation model for functional and logic programming. In *The 24th Symposium on Principles of Programming Languages*, pages 80–93, Paris, France, January 1997. ACM Press.

41. Michael Hanus and Frank Steiner. Controlling search in declarative programs. In Catuscia Palamidessi, Hugh Glaser, and Karl Meinke, editors, *Principles of Declarative Programming*, volume 1490 of *Lecture Notes in Computer Science*, pages 374–390, Pisa, Italy, September 1998. Springer-Verlag.

42. Seif Haridi and Per Brand. ANDORRA Prolog - an integration of Prolog and committed choice languages. In Institute for New Generation Computer Technology (ICOT), editor, *Proceedings of the International Conference on Fifth Generation Computer Systems*, volume 2, pages 745–754, Berlin, Germany, November 1988. Springer-Verlag.

43. Seif Haridi and Nils Franzén. *Tutorial of Oz*. The Mozart Consortium, www.mozart-oz.org, 1999.

44. Seif Haridi, Sverker Janson, and Catuscia Palamidessi. Structural operational semantics for AKL. *Future Generation Computer Systems*, 8:409–421, 1992.

45. Seif Haridi, Peter Van Roy, Per Brand, Michael Mehl, Ralf Scheidhauer, and Gert Smolka. Efficient logic variables for distributed computing. *ACM Transactions on Programming Languages and Systems*, 21(3):569–626, May 1999.

46. Seif Haridi, Peter Van Roy, Per Brand, and Christian Schulte. Programming languages for distributed applications. *New Generation Computing*, 16(3):223–261, 1998.

47. P[eter] E. Hart, N[ils] J. Nilsson, and B[ertram] Raphael. A formal basis for the heuristic determination of minimum cost paths. *IEEE Transactions on Systems Science and Cybernetics*, SSC-4(2):100–107, 1968.

48. William D. Harvey and Matthew L. Ginsberg. Limited discrepancy search. In Chris S. Mellish, editor, *Fourteenth International Joint Conference on Artificial Intelligence*, pages 607–615, Montréal, Québec, Canada, August 1995. Morgan Kaufmann Publishers.

49. Martin Henz. *Objects for Concurrent Constraint Programming*, volume 426 of *International Series in Engineering and Computer Science*. Kluwer Academic Publishers, Boston, MA, USA, October 1997.

50. Martin Henz. Constraint-based round robin tournament planning. In De Schreye [26], pages 545–557.

51. Martin Henz and Leif Kornstaedt. *The Oz Notation*. The Mozart Consortium, www.mozart-oz.org, 1999.

52. Martin Henz, Stefan Lauer, and Detlev Zimmermann.   COMPOzE — intention-based music composition through constraint programming. In *Proceedings of the 8th IEEE International Conference on Tools with Artificial Intelligence*, pages 118–121, Toulouse, France, November 1996. IEEE Computer Society Press.

53. Martin Henz, Tobias Müller, and Ka Boon Ng. Figaro: Yet another constraint programming library. In Inês de Castro Dutra, Vítor Santos Costa, Gopal Gupta, Enrico Pontelli, Manuel Carro, and Peter Kacsuk, editors, *Parallelism and Implementation Technology for (Constraint) Logic Programming*, pages 86–96, Las Cruces, NM, USA, December 1999. New Mexico State University.

54. Martin Henz and Jörg Würtz. Constraint-based time tabling—a case study. *Applied Artificial Intelligence*, 10(5):439–453, 1996.

55. ILOG S.A. *ILOG Solver 5.0: Reference Manual*. Gentilly, France, August 2000.

56. Joxan Jaffar, editor. *Proceedings of the Fifth International Conference on Principles and Practice of Constraint Programming*, volume 1713 of *Lecture Notes in Computer Science*. Springer-Verlag, Alexandra, VA, USA, October 1999.

57. Joxan Jaffar and Michael M. Maher. Constraint logic programming: A survey. *The Journal of Logic Programming*, 19 & 20:503–582, May 1994. Special Issue: Ten Years of Logic Programming.

58. Sverker Janson. *AKL - A Multiparadigm Programming Language*. PhD thesis, SICS Swedish Institute of Computer Science, SICS Box 1263, S-164 28 Kista, Sweden, 1994. SICS Dissertation Series 14.

59. Sverker Janson and Seif Haridi. Programming paradigms of the Andorra kernel language. In Vijay Saraswat and Kazunori Ueda, editors, *Logic Programming, Proceedings of the 1991 International Symposium*, pages 167–186, San Diego, CA, USA, October 1991. The MIT Press.

60. Sverker Janson, Johan Montelius, and Seif Haridi. Ports for objects. In *Research Directions in Concurrent Object-Oriented Programming*. The MIT Press, Cambridge, MA, USA, 1993.

61. Richard Jones and Rafael Lins. *Garbage Collection: Algorithms for Automatic Dynamic Memory Management*. John Wiley & Sons, Inc., New York, NY, USA, 1996.

62. Roland Karlsson. *A High Performance OR-parallel Prolog System*. PhD thesis, Swedish Institute of Computer Science, Kista, Sweden, March 1992.

63. Andrew J. Kennedy. Drawing trees. *Journal of Functional Programming*, 6(3):527–534, May 1996.

64. Donald E. Knuth. *The Art of Computer Programming*, volume 2 - Seminumerical Algorithms. Addison-Wesley, third edition, 1997.
65. Grzegorz Kondrak and Peter van Beek. A theoretical evaluation of selected backtracking algorithms. *Artificial Intelligence*, 89(1–2):365–187, January 1997.
66. Richard E. Korf. Depth-first iterative deepening: An optimal admissible tree search. *Artificial Intelligence*, 27(1):97–109, 1985.
67. Richard E. Korf. Optimal path-finding algorithms. In Laveen Kanal and Vipin Kumar, editors, *Search in Artificial Intelligence*, SYMBOLIC COMPUTATION – Artificial Intelligence, pages 223–267. Springer-Verlag, Berlin, Germany, 1988.
68. Richard E. Korf. Improved limited discrepancy search. In *Proceedings of the Thirteenth National Conference on Artificial Intelligence*, pages 286–291, Portland, OR, USA, August 1996. The MIT Press.
69. Robert Kowalski. *Logic for Problem Solving*. Elsevier Science Publishers, Amsterdam, The Netherlands, 1979.
70. Vipin Kumar and V. Nageshwara Rao. Parallel depth first search. Part II. Analysis. *International Journal of Parallel Programming*, 16(6):501–519, 1987.
71. François Laburthe. CHOCO: implementing a CP kernel. In Beldiceanu et al. [9], pages 71–85.
72. François Laburthe and Yves Caseau. SALSA: A language for search algorithms. In Maher and Puget [73], pages 310–324.
73. Michael Maher and Jean-François Puget, editors. *Proceedings of the Forth International Conference on Principles and Practice of Constraint Programming*, volume 1520 of *Lecture Notes in Computer Science*. Springer-Verlag, Pisa, Italy, October 1998.
74. Michael J. Maher. Logic semantics for a class of committed-choice programs. In Jean-Louis Lassez, editor, *Proceedings of the Fourth International Conference on Logic Programming*, pages 858–876, Melbourne, Australia, 1987. The MIT Press.
75. Kim Marriott and Peter J. Stuckey. *Programming with Constraints. An Introduction*. The MIT Press, Cambridge, MA, USA, 1998.
76. Michael Mehl. *The Oz Virtual Machine: Records, Transients, and Deep Guards*. Dissertation, Universität des Saarlandes, Im Stadtwald, 66041 Saarbrücken, Germany, 1999.
77. Michael Mehl, Ralf Scheidhauer, and Christian Schulte. An abstract machine for Oz. In Manuel Hermenegildo and S. Doaitse Swierstra, editors, *Programming Languages, Implementations, Logics and Programs, Seventh International Symposium, PLILP'95*, volume 982 of *Lecture Notes in Computer Science*, pages 151–168, Utrecht, The Netherlands, September 1995. Springer-Verlag.
78. Michael Mehl, Christian Schulte, and Gert Smolka. Futures and by-need synchronization for Oz. Draft, Programming Systems Lab, Universität des Saarlandes, May 1998.
79. Micha Meier. Debugging constraint programs. In Montanari and Rossi [88], pages 204–221.
80. Pedro Meseguer. Interleaved depth-first search. In Pollack [104], pages 1382–1387.
81. Pedro Meseguer and Toby Walsh. Interleaved and discrepancy based search. In Henri Prade, editor, *Proceedings of the 13th European Conference on Artificial Intelligence*, pages 239–243, Brighton, UK, 1998. John Wiley & Sons, Inc.

82. Tobias Müller. *Problem Solving with Finite Set Constraints.* The Mozart Consortium, www.mozart-oz.org, 1999.

83. Tobias Müller. Practical investigation of constraints with graph views. In Rina Dechter, editor, *Proceedings of the Sixth International Conference on Principles and Practice of Constraint Programming*, volume 1894 of *Lecture Notes in Computer Science*, pages 320–336, Singapore, September 2000. Springer-Verlag.

84. Tobias Müller. *Constraint Propagation in Mozart.* Dissertation, Universität des Saarlandes, Fakultät für Mathematik und Informatik, Fachrichtung Informatik, Im Stadtwald, 66041 Saarbrücken, Germany, 2001. In preparation.

85. Tobias Müller and Martin Müller. Finite set constraints in Oz. In François Bry, Burkhard Freitag, and Dietmar Seipel, editors, *13. Workshop Logische Programmierung*, pages 104–115, Technische Universität München, September 1997.

86. Tobias Müller and Jörg Würtz. Extending a concurrent constraint language by propagators. In Jan Małuszyński, editor, *Proceedings of the International Logic Programming Symposium*, pages 149–163, Long Island, NY, USA, 1997. The MIT Press.

87. Tobias Müller and Jörg Würtz. Embedding propagators in a concurrent constraint language. *The Journal of Functional and Logic Programming*, 1999(1):Article 8, April 1999. Special Issue. Available at: mitpress.mit.edu/JFLP/.

88. Ugo Montanari and Francesca Rossi, editors. *Proceedings of the First International Conference on Principles and Practice of Constraint Programming*, volume 976 of *Lecture Notes in Computer Science*. Springer-Verlag, Cassis, France, September 1995.

89. Johan Montelius. *Exploiting Fine-grain Parallelism in Concurrent Constraint Languages.* PhD thesis, SICS Swedish Institute of Computer Science, SICS Box 1263, S-164 28 Kista, Sweden, April 1997. SICS Dissertation Series 25.

90. Johan Montelius and Khayri A. M. Ali. An And/Or-parallel implementation of AKL. *New Generation Computing*, 13–14, August 1995.

91. Remco Moolenaar and Bart Demoen. Hybrid tree search in the Andorra model. In Van Hentenryck [141], pages 110–123.

92. Mozart Consortium. The Mozart programming system, 1999. Available from www.mozart-oz.org.

93. Shyam Mudambi and Joachim Schimpf. Parallel CLP on heterogeneous networks. In Van Hentenryck [141], pages 124–141.

94. J. F. Muth and G. L. Thompson. *Industrial Scheduling.* Prentice-Hall International, Englewood Cliffs, NY, USA, 1963.

95. Lee Naish. *Negation and Control in Prolog*, volume 238 of *Lecture Notes in Computer Science*. Springer-Verlag, Berlin, Germany, September 1986.

96. Lee Naish. Negation and quantifiers in NU-Prolog. In Ehud Shapiro, editor, *Proceedings of the Third International Conference on Logic Programming*, Lecture Notes in Computer Science, pages 624–634, London, UK, 1986. Springer-Verlag.

97. Nils J. Nilsson. *Principles of Artificial Intelligence.* Springer-Verlag, Berlin, Germany, 1982.

98. William Older and André Vellino. Constraint arithmetic on real intervals. In Benhamou and Colmerauer [10], pages 175–196.

99. John K. Ousterhout. *Tcl and the Tk Toolkit.* Professional Computing Series. Addison-Wesley, Reading, MA, USA, 1994.

100. Laurent Perron. Search procedures and parallelism in constraint programming. In Jaffar [56], pages 346–360.

101. Andreas Podelski, editor. *Constraint Programming: Basics and Trends*, volume 910 of *Lecture Notes in Computer Science*. Springer-Verlag, 1995.

102. Andreas Podelski and Gert Smolka. Situated simplification. *Theoretical Computer Science*, 173:209–233, February 1997.

103. Andreas Podelski and Peter Van Roy. The beauty and beast algorithm: quasi-linear incremental tests of entailment and disentailment over trees. In Bruynooghe [11], pages 359–374.

104. Martha E. Pollack, editor. *Fifteenth International Joint Conference on Artificial Intelligence*. Morgan Kaufmann Publishers, Nagoya, Japan, August 1997.

105. Enrico Pontelli and Vítor Santos Costa, editors. *Practical Aspects of Declarative Languages, Second International Workshop, PADL 2000*, volume 1753 of *Lecture Notes in Computer Science*. Springer-Verlag, Boston, MA, USA, January 2000.

106. Konstantin Popov. An implementation of distributed computation in programming language Oz. Master's thesis, Electrotechnical State University of St. Petersburg Uljanov/Lenin, February 1994. In Russian.

107. Konstantin Popov. A parallel abstract machine for the thread-based concurrent constraint language Oz. In Inês de Castro Dutra, Vítor Santos Costa, Fernando Silva, Enrico Pontelli, and Gopal Gupta, editors, *Workshop On Parallelism and Implementation Technology for (Constraint) Logic Programming Languages*, 1997.

108. Konstantin Popov. *The Oz Browser*. The Mozart Consortium, www.mozart-oz.org, 1999.

109. Steven Prestwich and Shyam Mudambi. Improved branch and bound in constraint logic programming. In Montanari and Rossi [88], pages 533–548.

110. Jean-François Puget. A C++ implementation of CLP. In *Proceedings of the Second Singapore International Conference on Intelligent Systems (SPICIS)*, pages B256–B261, Singapore, November 1994.

111. Jean-François Puget and Michel Leconte. Beyond the glass box: Constraints as objects. In John Lloyd, editor, *Proceedings of the International Symposium on Logic Programming*, pages 513–527, Portland, OR, USA, December 1995. The MIT Press.

112. Desh Ranjan, Enrico Pontelli, and Gopal Gupta. The complexity of or-parallelism. *New Generation Computing*, 17(3):285–308, 1999.

113. V. Nageshwara Rao and Vipin Kumar. Parallel depth first search. Part I. Implementation. *International Journal of Parallel Programming*, 16(6):479–499, 1987.

114. Stuart Russell and Peter Norvig. *Artificial Intelligence: A Modern Approach*. Prentice-Hall International, Englewood Cliffs, NY, USA, 1995.

115. Georg Sander. Graph layout through the VCG tool. In Tamassia and Tollis [139], pages 194–205.

116. Vítor Santos Costa, David H. D. Warren, and Rong Yang. The Andorra-I preprocessor: Supporting full Prolog on the Basic Andorra Model. In Koichi Furukawa, editor, *Proceedings of the Eight International Conference on Logic Programming*, pages 443–456, Paris, France, June 1991. The MIT Press.

117. Vijay A. Saraswat. *Concurrent Constraint Programming*. ACM Doctoral Dissertation Awards: Logic Programming. The MIT Press, Cambridge, MA, USA, 1993.

118. Vijay A. Saraswat and Martin Rinard. Concurrent constraint programming. In *Proceedings of the 7th Annual ACM Symposium on Principles of Programming Languages*, pages 232–245, San Francisco, CA, USA, January 1990. ACM Press.

119. Ralf Scheidhauer. *Design, Implementierung und Evaluierung einer virtuellen Maschine für Oz*. Dissertation, Universität des Saarlandes, Im Stadtwald, 66041 Saarbrücken, Germany, December 1998. In German.

120. Klaus Schild and Jörg Würtz. Off-line scheduling of a real-time system. In K. M. George, editor, *Proceedings of the 1998 ACM Symposium on Applied Computing, SAC98*, pages 29–38, Atlanta, GA, USA, 1998. ACM Press.

121. Christian Schulte. Oz Explorer: A visual constraint programming tool. In Lee Naish, editor, *Proceedings of the Fourteenth International Conference on Logic Programming*, pages 286–300, Leuven, Belgium, July 1997. The MIT Press.

122. Christian Schulte. Programming constraint inference engines. In Smolka [135], pages 519–533.

123. Christian Schulte. Comparing trailing and copying for constraint programming. In De Schreye [26], pages 275–289.

124. Christian Schulte. *Oz Explorer: Visual Constraint Programming Support*. The Mozart Consortium, www.mozart-oz.org, 1999.

125. Christian Schulte. *Window Programming in Mozart*. The Mozart Consortium, www.mozart-oz.org, 1999.

126. Christian Schulte. Parallel search made simple. In Beldiceanu et al. [9], pages 41–57.

127. Christian Schulte. Programming deep concurrent constraint combinators. In Pontelli and Costa [105], pages 215–229.

128. Christian Schulte and Gert Smolka. Encapsulated search in higher-order concurrent constraint programming. In Bruynooghe [11], pages 505–520.

129. Christian Schulte, Gert Smolka, and Jörg Würtz. Encapsulated search and constraint programming in Oz. In Alan H. Borning, editor, *Second Workshop on Principles and Practice of Constraint Programming*, volume 874 of *Lecture Notes in Computer Science*, pages 134–150, Orcas Island, WA, USA, May 1994. Springer-Verlag.

130. Helmut Simonis and Abder Aggoun. Search-tree visualization. In Deransart et al. [28], chapter 7, pages 191–208.

131. Jeffrey Mark Siskind and David Allen McAllester. Screamer: A portable efficient implementation of nondeterministic Common Lisp. Technical Report IRCS-93-03, University of Pennsylvania, Institute for Research in Cognitive Science, 1993.

132. Gert Smolka. Residuation and guarded rules for constraint logic programming. In Benhamou and Colmerauer [10], pages 405–419.

133. Gert Smolka. The definition of Kernel Oz. In Podelski [101], pages 251–292.

134. Gert Smolka. The Oz programming model. In Jan van Leeuwen, editor, *Computer Science Today*, volume 1000 of *Lecture Notes in Computer Science*, pages 324–343. Springer-Verlag, Berlin, 1995.

135. Gert Smolka, editor. *Proceedings of the Third International Conference on Principles and Practice of Constraint Programming*, volume 1330 of *Lecture Notes in Computer Science*. Springer-Verlag, Schloß Hagenberg, Linz, Austria, October 1997.

136. Gert Smolka. Concurrent constraint programming based on functional programming. In Chris Hankin, editor, *Programming Languages and Systems*, volume 1381 of *Lecture Notes in Computer Science*, pages 1–11, Lisbon, Portugal, March 1998. Springer-Verlag.

137. Gert Smolka and Ralf Treinen. Records for logic programming. *The Journal of Logic Programming*, 18(3):229–258, April 1994.

138. Jan Sundberg and Claes Svensson. MUSE TRACE: A graphic tracer for OR-parallel Prolog. SICS Technical Report T90003, Swedish Institute of Computer Science, Kista, Sweden, 1990.

139. Roberto Tamassia and Ioannis G. Tollis, editors. *Graph Drawing, DIMACS International Workshop GD'94*, volume 894 of *Lecture Notes in Computer Science*. Springer-Verlag, Princeton, USA, 1995.

140. Pascal Van Hentenryck. *Constraint Satisfaction in Logic Programming*. Programming Logic Series. The MIT Press, Cambridge, MA, USA, 1989.

141. Pascal Van Hentenryck, editor. *Proceedings of the Eleventh International Conference on Logic Programming*. The MIT Press, Santa Margherita Ligure, Italy, 1994.

142. Pascal Van Hentenryck. *The OPL Optimization Programming Language*. The MIT Press, Cambridge, MA, USA, 1999.

143. Pascal Van Hentenryck and Yves Deville. The cardinality operator: A new logical connective for constraint logic programming. In Benhamou and Colmerauer [10], pages 383–403.

144. Pascal Van Hentenryck, Laurent Perron, and Jean-François Puget. Search and strategies in OPL. *ACM Transactions on Computational Logic*, 1(2):285–320, October 2000.

145. Pascal Van Hentenryck, Vijay Saraswat, and Yves Deville. Design, implementation, and evaluation of the constraint language cc(FD). *The Journal of Logic Programming*, 37(1–3):139–164, October 1998.

146. Pascal Van Hentenryck, Vijay Saraswat, et al. Strategic directions in constraint programming. *ACM Computing Surveys*, 28(4):701–726, December 1997. ACM 50th Anniversary Issue. Strategic Directions in Computing Research.

147. Peter Van Roy, Seif Haridi, and Per Brand. *Distributed Programming in Mozart - A Tutorial Introduction*. The Mozart Consortium, www.mozart-oz.org, 1999.

148. Peter Van Roy, Seif Haridi, Per Brand, Gert Smolka, Michael Mehl, and Ralf Scheidhauer. Mobile objects in Distributed Oz. *ACM Transactions on Programming Languages and Systems*, 19(5):804–851, September 1997.

149. Peter Van Roy, Michael Mehl, and Ralf Scheidhauer. Integrating efficient records into concurrent constraint programming. In Herbert Kuchen and S. Doaitse Swierstra, editors, *Programming Languages: Implementations, Logics, and Programs, 8th International Symposium, PLILP'96*, volume 1140 of *Lecture Notes in Computer Science*, pages 438–453, Aachen, Germany, September 1996. Springer-Verlag.

150. Mark Wallace. Practical applications of Constraint Programming. *Constraints*, 1(1/2):139–168, September 1996.

151. Mark Wallace, Stefano Novello, and Joachim Schimpf. Eclipse: A platform for constraint logic programming. Technical report, IC-Parc, Imperial College, London, GB, August 1997.

152. Joachim Paul Walser. Feasible cellular frequency assignment using constraint programming abstractions. In *Proceedings of the Workshop on Constraint Programming Applications, in conjunction with the Second International Conference on Principles and Practice of Constraint Programming (CP96)*, Cambridge, MA, USA, August 1996.

153. Toby Walsh. Depth-bounded discrepancy search. In Pollack [104], pages 1388–1393.

154. David H. D. Warren. An abstract Prolog instruction set. Technical Note 309, SRI International, Artificial Intelligence Center, Menlo Park, CA, USA, October 1983.

155. Paul R. Wilson. Uniprocessor garbage collection techniques. In Yves Bekkers and Jacques Cohen, editors, *Memory Management: International Workshop IWMM 92*, volume 637 of *Lecture Notes in Computer Science*, pages 1–42, St. Malo, France, September 1992. Springer-Verlag. Revised version will appear in ACM Computing Surveys.

156. Jörg Würtz. Constraint-based scheduling in Oz. In Uwe Zimmermann, Ulrich Derigs, Wolfgang Gaul, Rolf H. Möhring, and Karl-Peter Schuster, editors, *Operations Research Proceedings 1996*, pages 218–223, Braunschweig, Germany, 1997. Springer-Verlag.

157. Jörg Würtz. *Lösen von kombinatorischen Problemen mit Constraintprogrammierung in Oz*. Dissertation, Universität des Saarlandes, Fachbereich Informatik, Im Stadtwald, 66041 Saarbrücken, Germany, January 1998. In German.

Index

Lecture Notes in Artificial Intelligence (LNAI)

Lecture Notes in Computer Science